BUILDING SUSTAINABILITY

PRACTICAL CONSTRUCTION PRACTICES

Steven Smith

Wisdom Publishers

ISBN: 9798851429125
Imprint: Independently published

Cover design by: Art Painter
Library of Congress Control Number: 2018675309
Printed in the United States of America

To all the construction professionals, project managers, contractors, architects, and students who strive to build a better future. Your commitment to sustainable construction practices is an inspiration. May this book serve as a guide and empower you to create environmentally responsible, resilient, and innovative buildings that shape a greener world.

The future of construction is green, sustainable, and resilient. By embracing sustainable practices, we can build a better world for generations to come

UNITED NATIONS

CONTENTS

INTRODUCTION

In today's rapidly evolving world, the concept of sustainable construction has gained significant attention and importance within the construction industry. As we face pressing environmental challenges, such as climate change and resource depletion, there is a growing recognition that traditional construction practices must be reimagined to ensure a more sustainable and resilient future.

Sustainable construction encompasses an approach that seeks to minimize the negative environmental impact of buildings, while also addressing social and economic considerations. It goes beyond the mere construction of structures and involves integrating principles and practices that promote environmental stewardship, resource efficiency, and social responsibility throughout the entire lifecycle of a building.

Defining sustainable construction involves recognizing the interconnectedness of environmental, economic, and social factors. It embraces the idea that construction projects should not only be environmentally responsible but also economically viable and socially equitable. By adopting sustainable construction practices, we can create buildings that not only meet the needs of the present but also contribute to a sustainable and resilient future.

The importance of sustainable construction cannot be

overstated. It offers numerous benefits that span environmental, economic, and social dimensions. By reducing carbon emissions, conserving resources, and minimizing waste generation, sustainable construction mitigates the impacts of climate change and contributes to a healthier planet. Moreover, it enhances energy and cost efficiency, reduces operational expenses, and stimulates economic growth in the green construction sector. Socially, sustainable construction improves occupant health and well-being, promotes inclusivity and accessibility, and fosters stronger, more cohesive communities.

The benefits of sustainable construction practices are far-reaching. Environmentally, they lead to reduced energy consumption, water conservation, and preservation of natural resources. Economically, they result in long-term cost savings, increased property value, and enhanced marketability. Socially, they create healthier and more comfortable living and working environments, promote social equity, and support community well-being. By adopting sustainable construction practices, we not only address the immediate needs of a project but also contribute to a more sustainable and prosperous future.

The current trends in sustainable construction reflect the growing market demand for environmentally responsible and socially conscious buildings. Innovations in green design, advanced technologies, and digital solutions are reshaping the construction industry. Building certifications and standards, such as LEED, BREEAM, and WELL, have gained prominence, highlighting the importance of sustainability in meeting market expectations. Additionally, changing regulations, government initiatives, and industry collaborations are influencing the shift towards sustainable construction practices.

Understanding sustainable construction and its significance is crucial in addressing the environmental challenges we face. By recognizing the definition and importance of sustainable construction, recognizing its benefits, and staying informed about current trends and market demand, we can contribute to a more sustainable and resilient built environment. This book aims to delve deeper into these topics, providing construction professionals, project managers, contractors, architects, and students with the knowledge and insights needed to adopt and implement sustainable construction practices effectively.

PART I: SUSTAINABLE DESIGN PRINCIPLES

CHAPTER 1: INTEGRATING SUSTAINABILITY INTO DESIGN

1.1   *Sustainable Design Strategies and Approaches*

Sustainable design strategies and approaches form the bedrock of sustainable construction practices. By integrating environmental, economic, and social considerations into the design process, professionals can create buildings that are not only environmentally responsible but also promote occupant health, well-being, and productivity. In this section, we will explore key sustainable design strategies and approaches, examining their principles and benefits.

Holistic Design Thinking

Sustainable design begins with a holistic approach that considers the interdependencies between various elements of a building and its surrounding environment. Rather than focusing solely on individual components, holistic design thinking takes into account the entire lifecycle of the building, from construction and operation to maintenance and eventual renovation or demolition.

This approach requires collaboration between architects, engineers, contractors, and other stakeholders, ensuring that sustainability considerations are integrated into every decision-making process. By fostering interdisciplinary collaboration, designers can maximize the potential for sustainable outcomes and create buildings that are in harmony with their surroundings.

Passive Design Techniques

Passive design techniques aim to optimize a building's performance by capitalizing on natural resources and climatic conditions. By carefully analyzing the site, orientation, and climate, designers can utilize passive strategies to minimize energy consumption and enhance occupant comfort.

Orientation: Proper orientation of a building can harness solar energy and natural daylight, reducing the need for artificial lighting and heating. By maximizing the use of natural light and heat gain, designers can decrease reliance on energy-consuming systems.

Insulation and Thermal Mass: Effective insulation and thermal mass management play crucial roles in maintaining a comfortable indoor environment. Proper insulation reduces heat transfer, while thermal mass absorbs and releases heat, stabilizing temperature fluctuations.

Natural Ventilation: Incorporating natural ventilation strategies allows for the flow of fresh air, reducing reliance on mechanical cooling systems. This can be achieved through well-placed windows, vents, and building design that facilitates air circulation.

Energy-Efficient Design

Energy-efficient design focuses on minimizing energy consumption throughout the building's lifecycle. This involves implementing energy-efficient technologies and systems that reduce reliance on non-renewable energy sources.

Building Envelope: A well-designed building envelope, including insulation, air sealing, and high-performance windows, helps maintain a comfortable indoor environment while minimizing energy loss.

Efficient Lighting: Utilizing energy-efficient lighting systems, such as LED technology, reduces electricity consumption and enhances lighting quality. Incorporating daylight harvesting techniques further reduces the need for artificial lighting during daylight hours.

Energy-Efficient HVAC Systems: Heating, ventilation, and air conditioning (HVAC) systems account for a significant portion of a building's energy consumption. By incorporating energy-efficient HVAC technologies, such as variable refrigerant flow (VRF) systems and heat recovery ventilation, designers can optimize energy use while maintaining thermal comfort.

Renewable Energy Integration: Integrating renewable energy sources, such as solar panels or wind turbines, into the building's design allows for on-site generation of clean energy, reducing reliance on fossil fuel-based power.

Water Efficiency

Water efficiency in sustainable design involves reducing water consumption, optimizing water management, and promoting water reuse. By implementing water-efficient strategies, designers can contribute to water conservation efforts and reduce the strain on local water resources.

Low-Flow Fixtures: Installing low-flow faucets, showers, and toilets can significantly reduce water usage without compromising functionality or comfort.

Rainwater Harvesting: Collecting and storing rainwater for non-potable uses, such as irrigation or toilet flushing, reduces the demand for freshwater resources.

Greywater Systems: Greywater systems treat and reuse water from sinks, showers, and laundry for non-potable purposes, further reducing freshwater consumption.

Landscape Design: Implementing drought-tolerant landscaping and efficient irrigation techniques minimizes the water requirements of outdoor spaces.

Material Selection and Life Cycle Assessment

Sustainable design involves conscious material selection and a thorough assessment of a building's environmental impact throughout its life cycle. By considering factors such as resource extraction, manufacturing processes, embodied energy, and end-of-life disposal, designers can minimize the environmental footprint of a building.

Sustainable Materials: Opting for environmentally friendly materials, such as recycled or locally sourced materials, reduces the ecological impact associated with extraction, transportation, and manufacturing.

Life Cycle Assessment (LCA): Conducting a life cycle assessment helps evaluate the environmental impact of materials and systems over their entire life cycle. This assessment considers aspects such as energy use, greenhouse gas emissions, and waste generation, allowing designers to make informed decisions.

Design for Disassembly: Designing for disassembly promotes the ease of separating and recycling building components at the end of a building's life. This approach reduces waste and encourages the reuse of materials in future construction projects.

Professionals can leverage sustainable design strategies to create buildings that minimize resource consumption, reduce environmental impact, and provide healthier and more comfortable spaces for occupants. The adoption of these strategies not only aligns with sustainability goals but also contributes to long-term cost savings and societal well-being. As the construction industry continues to embrace sustainable practices, the integration of sustainable design principles becomes crucial for a greener and more sustainable future.

1.2   Passive Design Techniques for Energy Efficiency

Passive design techniques play a crucial role in creating energy-efficient buildings by harnessing natural resources and climatic conditions. By leveraging the principles of passive design, architects and engineers can significantly reduce the energy demand for heating, cooling, and lighting, resulting in lower energy consumption, reduced environmental impact, and increased occupant comfort. In this section, we will delve into the key passive design techniques and explore their implementation and benefits in achieving energy-efficient buildings.

Orientation and Building Form

The orientation and form of a building are fundamental considerations in passive design. By strategically positioning a building and optimizing its shape, designers can maximize exposure to beneficial solar radiation while minimizing unwanted heat gain or loss.

South-facing Orientation: In the Northern Hemisphere, maximizing the south-facing facade allows for optimal solar access during winter months when the sun is lower in the sky. This allows the building to capture passive solar heat gain, reducing the need for artificial heating.

Shading and Overhangs: Incorporating shading devices, such as roof overhangs, louvers, or awnings, can prevent excessive solar heat gain during warmer months. These elements block direct sunlight from entering the building while still allowing natural light to penetrate.

Building Shape: The form and layout of a building influence its exposure to wind and natural ventilation. Designing buildings with compact shapes or incorporating wind-responsive design features can minimize heat loss through drafts and promote effective natural ventilation.

Building Envelope and Insulation

The building envelope, consisting of walls, roofs, floors, and windows, plays a critical role in controlling heat transfer and maintaining thermal comfort within a building. Proper insulation and thermal resistance help reduce heat gain in warmer months and heat loss during colder periods.

Insulation: High-quality insulation materials, such as mineral wool, cellulose, or foam insulation, effectively minimize heat transfer through walls, roofs, and floors. Insulation with high thermal resistance helps maintain comfortable indoor temperatures and reduces the need for heating or cooling.

Thermal Mass: Incorporating thermal mass materials, such as concrete or masonry, can help stabilize indoor temperatures by absorbing and releasing heat slowly. This moderates temperature fluctuations, reducing the need for mechanical heating or cooling.

Air Sealing: Effective air sealing of the building envelope prevents air leakage, minimizing energy loss and improving energy efficiency. Properly sealing gaps, cracks, and joints ensures airtightness and enhances the performance of insulation and HVAC systems.

Natural Ventilation

Natural ventilation refers to the intentional use of natural airflow to cool and ventilate a building. By incorporating design elements that promote natural ventilation, designers can reduce reliance on mechanical cooling systems, improving energy efficiency and indoor air quality.

Building Orientation and Layout: The layout and orientation of spaces within a building can be designed to encourage natural airflow. Utilizing prevailing winds and designing cross-ventilation paths allow for the flow of fresh air, promoting cooling and air exchange.

Operable Windows and Vents: Incorporating operable windows and vents enables occupants to control airflow and take advantage of natural breezes. Designing windows and vents that can be easily opened or closed allows for adaptable ventilation based on outdoor conditions.

Stack Effect: Utilizing the stack effect, which leverages the natural buoyancy of warm air, can facilitate airflow and cooling within a building. Warm air rises, creating a pressure difference that draws cooler air from lower openings, promoting natural ventilation.

Daylighting
Maximizing natural daylight not only reduces the need for artificial lighting but also creates visually pleasing and productive spaces. Proper daylighting design enhances occupant well-being, comfort, and energy efficiency.

Window Placement and Size: Optimizing window placement and size allows for the penetration of natural light deep into interior spaces. Placing windows strategically to capture daylight while avoiding excessive heat gain or glare is crucial.

Light Shelves and Reflective Surfaces: Light shelves and reflective surfaces can be incorporated to redirect and distribute natural light deeper into a space. These elements bounce daylight off ceilings and walls, reducing the need for artificial lighting.

Light Control and Glare Reduction: Utilizing shading devices, such as blinds or solar control films, can help regulate incoming daylight and reduce glare. Balancing the amount of daylight entering a space ensures visual comfort while minimizing heat gain.

Natural Landscaping

The surrounding landscape can contribute to the energy efficiency of a building. Careful planning and design of outdoor spaces can provide shading, reduce heat island effects, and contribute to overall thermal comfort.

Deciduous Trees and Vegetation: Planting deciduous trees strategically around a building provides shade during the summer months when leaves are abundant, while allowing sunlight to reach the building during the winter months when leaves have shed. This natural shading reduces cooling demands.

Green Roofs and Walls: Incorporating green roofs and walls can improve energy efficiency by providing additional insulation, reducing heat island effects, and enhancing stormwater management. Vegetation on roofs and walls absorbs solar radiation, lowering the temperature and reducing cooling needs.

Water Features: Introducing water features, such as fountains or ponds, can contribute to cooling the surrounding environment through evaporation. The evaporation process absorbs heat, lowering temperatures and improving comfort.

Architects and engineers can achieve significant energy savings and create buildings that provide comfortable, healthy, and sustainable environments for occupants. These techniques work synergistically to optimize energy performance, reduce reliance on mechanical systems, and minimize environmental impact. Incorporating passive design strategies into building projects not only benefits the occupants but also contributes to a greener and more sustainable future.

1.3 Water Conservation and Management in Design

Water conservation and management play crucial roles in sustainable construction practices. With increasing concerns over water scarcity and the need to protect this precious resource, incorporating effective water conservation strategies into the design of buildings becomes essential. By implementing water-efficient systems and practices, designers can significantly reduce water consumption, minimize the strain on local water supplies, and contribute to a more sustainable and resilient built environment. In this section, we will explore key aspects of water conservation and management in design, examining strategies, technologies, and benefits.

Efficient Plumbing Fixtures and Appliances

One of the primary areas for water conservation in building design lies in the selection and installation of efficient plumbing fixtures and appliances. By utilizing low-flow fixtures and water-saving technologies, designers can significantly reduce water consumption without compromising functionality or user experience.

Low-Flow Faucets and Showerheads: Installing low-flow faucets and showerheads helps reduce water flow rates while maintaining satisfactory water pressure and performance. These fixtures incorporate aerators or flow restrictors that mix air with water, delivering a satisfying flow while conserving water.

Water-Efficient Toilets: Water-efficient toilets, such as dual-flush or low-flush models, reduce the volume of water used for flushing. Dual-flush toilets provide different flush options for solid and liquid waste, allowing users to select the appropriate water volume needed.

Waterless Urinals: Waterless urinals eliminate the need for flushing, resulting in significant water savings. These urinals use specialized mechanisms that rely on gravity and sealants to prevent odor and maintain hygiene.

Greywater Systems

Greywater systems enable the collection and treatment of non-potable water from sources such as sinks, showers, and laundry. By reusing greywater for irrigation or non-potable uses, designers can further reduce the demand for freshwater and conserve water resources.

Greywater Treatment and Filtration: Greywater treatment systems involve processes like filtration, sedimentation, and disinfection to remove impurities and pathogens. Treated greywater can then be safely reused for flushing toilets, irrigation, or other non-potable applications.

Separate Plumbing Systems: Designing buildings with separate plumbing systems for greywater and freshwater allows for efficient collection, treatment, and distribution of greywater. This ensures that the treated greywater is appropriately isolated from the freshwater supply.

Rainwater Harvesting

Rainwater harvesting involves the collection and storage of rainwater for future use. By capturing and utilizing rainwater, designers can supplement freshwater supply and reduce the strain on municipal water systems.

Roof and Surface Collection: Designing buildings with roof gutters, downspouts, and surface drainage systems enables the collection of rainwater from rooftops and other surfaces. This water can be directed to storage tanks or reservoirs for later use.

Filtration and Treatment: Rainwater should undergo appropriate filtration and treatment before use, particularly for potable purposes. Filtration systems remove debris and contaminants, while disinfection processes ensure the water's safety and quality.

Non-Potable Uses: Harvested rainwater can be utilized for non-potable applications such as landscape irrigation, toilet flushing, or cleaning. By utilizing rainwater for these purposes, designers can reduce the demand for freshwater and conserve water resources.

Efficient Irrigation Practices

Efficient irrigation practices are essential for minimizing water consumption in outdoor spaces. By adopting water-efficient irrigation systems and landscaping strategies, designers can reduce water waste and promote sustainable water management.

Drip Irrigation: Drip irrigation systems deliver water directly to plant roots, minimizing water loss due to evaporation or runoff. This targeted approach ensures efficient water distribution, reducing overall water consumption.

Smart Irrigation Controllers: Smart controllers utilize weather data, soil moisture sensors, and plant water needs to optimize irrigation schedules and water delivery. By adjusting irrigation based on real-time conditions, smart controllers prevent overwatering and promote water efficiency.

Xeriscaping and Native Plant Selection: Xeriscaping involves selecting drought-tolerant plants and utilizing landscaping techniques that reduce water requirements. Native plants, adapted to the local climate, generally require less water and maintenance, reducing the demand for irrigation.

Stormwater Management

Effective stormwater management is crucial for mitigating the impact of rainfall events and reducing strain on municipal drainage systems. By implementing sustainable stormwater management practices, designers can minimize runoff and enhance water infiltration.

Permeable Surfaces: Designing permeable surfaces, such as pervious pavements or green spaces, allows rainwater to infiltrate the ground, reducing runoff and replenishing groundwater resources. This approach also helps filter out pollutants and improve water quality.

Rain Gardens and Bioswales: Rain gardens and bioswales are landscape features designed to capture and manage stormwater runoff. These vegetated areas collect and naturally filter runoff, promoting water infiltration and reducing erosion.

Detention and Retention Systems: Detention and retention systems, such as rainwater harvesting tanks or underground storage chambers, collect and store stormwater for controlled release or later use. These systems help manage peak flows and prevent flooding while conserving water for future needs.

By incorporating water conservation and management strategies into building design, designers can make a significant impact on water conservation efforts and contribute to a more sustainable future. These strategies not only reduce water consumption but also promote responsible water management, alleviate stress on water resources, and enhance overall resilience in the face of water scarcity challenges.

The benefits of incorporating water conservation and management in design are multifaceted. Firstly, it reduces the strain on municipal water supply systems, particularly in regions prone to water scarcity or experiencing population growth. By minimizing water demand, buildings equipped with efficient plumbing fixtures, greywater systems, and rainwater harvesting systems help alleviate the burden on local water infrastructure, ensuring a more reliable and sustainable water supply for all.

Secondly, water conservation in design leads to significant cost savings. By reducing water consumption, building owners can experience reduced water bills and operational expenses. Additionally, the implementation of water-efficient systems and practices can contribute to meeting sustainability goals and obtaining green building certifications, such as LEED (Leadership in Energy and Environmental Design), which further enhances the market value and marketability of the building.

Moreover, water conservation practices contribute to environmental sustainability. By conserving water resources, designers reduce the need for large-scale water extraction, thereby minimizing the ecological impact on natural water sources. Additionally, the reduction of water consumption reduces the energy required for water treatment and distribution, resulting in reduced greenhouse gas emissions and a lower carbon footprint associated with water infrastructure.

Finally, incorporating water conservation and management strategies aligns with societal expectations and promotes a culture of responsible water stewardship. Buildings designed with water-efficient features demonstrate a commitment to environmental sustainability and inspire others to adopt similar practices. By setting an example and raising awareness of the importance of water conservation, designers contribute to a broader movement towards sustainable water management practices.

Water conservation and management in design are critical elements of sustainable construction practices. By implementing efficient plumbing fixtures, greywater systems, rainwater harvesting, and adopting water-efficient irrigation and stormwater management practices, designers can minimize water consumption, promote responsible water use, and contribute to a more sustainable built environment. The integration of these strategies not only benefits the environment and local communities but also leads to cost savings, enhances market value, and helps meet sustainability goals. Through collective efforts, designers can play a vital role in addressing water scarcity challenges and shaping a more water-conscious and sustainable future.

CHAPTER 2: MATERIALS SELECTION FOR SUSTAINABILITY

2.1 Life Cycle Assessment of Building Materials

The life cycle assessment (LCA) of building materials is a comprehensive methodology that evaluates the environmental impact of materials throughout their entire life cycle - from extraction and production to use, maintenance, and disposal. By conducting a thorough LCA, designers can make informed decisions about material selection and identify opportunities to reduce the environmental footprint of buildings. In this section, we will delve into the details of LCA for building materials, examining its stages, key considerations, and benefits.

Stages of Life Cycle Assessment

The life cycle assessment of building materials typically consists of four stages:

Raw Material Acquisition: This stage involves assessing the environmental impact associated with the extraction, processing, and transportation of raw materials used in building materials. It considers factors such as energy consumption, land use, water usage, and emissions resulting from extraction activities.

Manufacturing and Production: During this stage, the energy consumption, emissions, waste generation, and other environmental impacts associated with the manufacturing and production processes of building materials are evaluated. This includes processes such as heating, cooling, chemical reactions, and material transformation.

Use and Maintenance: The use and maintenance stage assesses the environmental impact of building materials during their service life. It considers factors such as energy consumption, water usage, and emissions associated with the use, operation, and maintenance of the building materials and components.

End-of-Life Disposal: The end-of-life stage examines the environmental impact of building material disposal, including recycling, reuse, or waste management processes. It evaluates factors such as waste generation, emissions from incineration or landfilling, and the potential for material recovery or recycling.

Key Considerations in Life Cycle Assessment

When conducting a life cycle assessment of building materials, several key considerations should be taken into account:

Embodied Energy: Embodied energy refers to the energy consumed throughout the life cycle of a building material, including the energy required for extraction, processing, manufacturing, transportation, and disposal. Assessing embodied energy helps identify materials with lower energy demands and reduce the overall environmental impact of a building.

Global Warming Potential: Global warming potential measures the greenhouse gas emissions associated with building materials. It evaluates emissions of carbon dioxide (CO_2), methane (CH_4), nitrous oxide (N_2O), and other greenhouse gases. By considering the global warming potential of materials, designers can select low-carbon alternatives and contribute to climate change mitigation efforts.

Resource Depletion: The assessment should consider the depletion of non-renewable resources, such as fossil fuels, minerals, and metals, used in the production of building materials. Identifying materials with lower resource depletion rates or utilizing recycled and renewable materials can help reduce the strain on finite resources.

Water Usage: Water consumption during the life cycle of building materials is an essential consideration. Assessing water usage helps identify materials and production processes with lower water demands, contributing to water conservation efforts and reducing the ecological impact on water resources.

Pollution and Emissions: LCA evaluates the emissions of pollutants, such as volatile organic compounds (VOCs), particulate matter, and hazardous substances, associated with building materials. By minimizing emissions and pollution, designers can create healthier indoor and outdoor environments for building occupants and the surrounding community.

Benefits of Life Cycle Assessment

Life cycle assessment provides numerous benefits for designers, construction professionals, and stakeholders involved in building projects:

Informed Material Selection: LCA enables designers to make informed decisions regarding material selection. By considering the environmental impact of materials across their life cycle, designers can identify environmentally preferable alternatives, reducing the overall ecological footprint of a building.

Environmental Performance Evaluation: LCA provides a comprehensive assessment of a building's environmental performance. It allows designers to quantify the environmental impacts associated with different materials and production processes, enabling targeted improvements and optimizing sustainability performance.

Sustainable Certifications and Standards: LCA serves as a valuable tool for achieving sustainable certifications and complying with green building standards. Certifications such as LEED (Leadership in Energy and Environmental Design) and BREEAM (Building Research Establishment Environmental Assessment Method) require the consideration of LCA as part of their assessment criteria, encouraging the use of environmentally responsible materials.

Cost Reduction and Resource Efficiency: LCA helps identify opportunities for cost reduction and resource efficiency. By selecting materials with lower embodied energy or utilizing recycled and renewable materials, designers can potentially reduce energy consumption, waste generation, and material costs, leading to economic savings.

Improved Environmental Reputation: Incorporating LCA into the design process demonstrates a commitment to environmental responsibility and sustainability. It enhances the reputation of designers, construction professionals, and building owners by showcasing their efforts in minimizing environmental impact and contributing to a greener future.

Designers should consider life cycle assessments of building materials to promote sustainable and responsible material selection, optimize environmental performance, and contribute to the overall sustainability of buildings. LCA provides a robust methodology for evaluating the environmental impacts associated with materials, guiding decision-making processes, and facilitating the transition to a more sustainable construction industry.

2.2 Sustainable Sourcing and Material Selection

Sustainable sourcing and material selection are crucial components of environmentally responsible construction practices. By carefully considering the origin, production processes, and environmental impact of building materials, designers can make informed decisions that minimize the ecological footprint of construction projects. In this section, we will explore the importance of sustainable sourcing and material selection, key considerations, and strategies for promoting sustainable practices in the construction industry.

Importance of Sustainable Sourcing

Sustainable sourcing refers to the responsible procurement of building materials with consideration for environmental, social, and economic factors. It involves evaluating the entire supply chain, from raw material extraction to manufacturing, transportation, and distribution. Adopting sustainable sourcing practices is crucial for the following reasons:

Environmental Impact: Unsustainable sourcing practices can contribute to deforestation, habitat destruction, soil erosion, water pollution, and greenhouse gas emissions. By prioritizing sustainable sourcing, designers can reduce the ecological impact associated with material extraction and production, helping to preserve ecosystems, conserve natural resources, and mitigate climate change.

Social Responsibility: Sustainable sourcing promotes fair labor practices, worker safety, and social equity. It encourages responsible supply chain management, ensuring that workers involved in material extraction and manufacturing are treated ethically and provided with safe working conditions. By choosing suppliers that prioritize social responsibility, designers can support equitable and socially responsible practices throughout the industry.

Health and Well-being: Some building materials can contain harmful substances that pose risks to human health and well-being. Sustainable sourcing emphasizes the selection of materials with low toxicity and emissions, promoting healthier indoor environments for occupants. By avoiding materials with harmful chemicals, designers can enhance the overall quality and safety of buildings.

Market Transformation: Embracing sustainable sourcing practices can drive market transformation by creating demand for environmentally responsible materials. By selecting sustainably sourced materials, designers send a clear message to suppliers, manufacturers, and the industry as a whole that sustainability is a priority. This stimulates innovation and encourages the development of greener alternatives, ultimately shifting the entire market toward more sustainable practices.

Key Considerations in Material Selection

When selecting building materials, several key considerations should be taken into account to ensure sustainability:

Life Cycle Assessment (LCA): Life cycle assessment evaluates the environmental impact of materials throughout their entire life cycle. It considers factors such as resource extraction, manufacturing processes, energy consumption, waste generation, and end-of-life disposal. By analyzing the life cycle of materials, designers can identify options with lower environmental footprints and make informed decisions.

Renewable and Recycled Content: Choosing materials with renewable or recycled content helps reduce reliance on virgin resources and promotes the circular economy. Materials made from renewable resources, such as sustainably harvested wood or bamboo, have a lower environmental impact compared to those derived from non-renewable sources. Similarly, utilizing recycled materials reduces waste generation and the need for resource extraction.

Durability and Longevity: Opting for durable materials that can withstand the test of time reduces the need for frequent replacements and renovations. Long-lasting materials contribute to lower maintenance requirements, reduced waste generation, and lower life cycle costs. It is essential to consider the durability and lifespan of materials when making selections.

Energy Efficiency: Energy-efficient materials contribute to reduced energy consumption during the operational phase of a building. For example, selecting high-performance insulation or energy-efficient windows can minimize heating and cooling demands. By prioritizing energy-efficient materials, designers can improve the overall energy performance of buildings and reduce greenhouse gas emissions.

Local Sourcing: Choosing locally sourced materials reduces transportation distances, lowering carbon emissions associated with shipping and logistics. Additionally, supporting local suppliers and manufacturers stimulates the local economy and fosters community resilience. Local sourcing can also promote the use of regionally available materials, which are often more suitable for the local climate and conditions.

Certification and Standards: Building certifications and standards, such as Forest Stewardship Council (FSC) for wood products or Cradle to Cradle (C2C) for overall sustainability, provide guidelines and verification of sustainable material sourcing and production. These certifications ensure that materials meet specific sustainability criteria, enabling designers to make reliable choices.

Strategies for Promoting Sustainable Sourcing

To promote sustainable sourcing and material selection, designers can adopt the following strategies:

Collaboration and Communication: Effective communication and collaboration with suppliers, manufacturers, and industry partners are key to promoting sustainable sourcing. Engaging in dialogue about sustainability goals and expectations encourages suppliers to adopt sustainable practices and provides designers with access to information about materials' environmental credentials.

Supplier Assessments: Conducting thorough assessments of suppliers' sustainability practices helps ensure that materials are sourced from responsible and environmentally conscious sources. Assessments may include evaluating suppliers' environmental policies, certifications, and compliance with relevant standards.

Material Databases and Tools: Utilizing material databases and assessment tools, such as the Building Research Establishment Environmental Assessment Method (BREEAM) or the Environmental Product Declarations (EPDs) database, can provide valuable information on the environmental performance of materials. These resources assist designers in making sustainable material choices based on reliable data and verified assessments.

Education and Training: Building awareness and knowledge about sustainable sourcing practices within the industry is crucial. By providing education and training to designers, contractors, and other stakeholders, the understanding and demand for sustainable materials can be enhanced, leading to widespread adoption of sustainable sourcing practices.

Continuous Improvement: Sustainable sourcing is an ongoing process that requires continuous improvement. Regularly reviewing material selections, staying updated on advancements in sustainable materials, and monitoring suppliers' sustainability practices contribute to ongoing progress and ensure that sustainable sourcing practices are consistently integrated into construction projects.

Sustainable sourcing and material selection are vital aspects of environmentally responsible construction practices. By prioritizing sustainable sourcing, considering life cycle impacts, and selecting materials with low environmental footprints, designers can contribute to a more sustainable built environment. Implementing strategies such as collaboration, supplier assessments, and utilizing material databases support the adoption of sustainable practices throughout the industry. Through conscious material choices, designers can reduce the environmental impact of construction projects, support social responsibility, and drive market transformation towards a more sustainable future.

2.3 Green Building Materials and Technologies

Green building materials and technologies are essential components of sustainable construction practices. By choosing environmentally friendly materials and integrating innovative technologies, designers can create buildings that minimize their environmental impact, promote energy efficiency, and enhance occupant health and well-being. In this section, we will explore the concept of green building materials and technologies, examine their benefits, and highlight key examples in the construction industry.

Understanding Green Building Materials

Green building materials are products and systems that are designed and manufactured with a focus on environmental sustainability. These materials prioritize resource efficiency, energy conservation, and the reduction of harmful emissions. They are typically sourced, produced, used, and disposed of in a manner that minimizes negative environmental impacts and promotes long-term sustainability.

Renewable Materials: Renewable materials are derived from natural resources that can be replenished over time, either through natural processes or human intervention. Examples include sustainably harvested wood, bamboo, cork, and natural fibers. Utilizing renewable materials reduces reliance on non-renewable resources and supports the principles of the circular economy.

Recycled and Upcycled Materials: Recycled and upcycled materials are produced by processing and transforming waste or byproducts into new building products. These materials divert waste from landfills and reduce the demand for virgin resources. Examples include recycled steel, reclaimed wood, recycled glass, and upcycled plastics.

Low-Energy and Low-Impact Materials: Low-energy materials are manufactured with reduced energy consumption, often through innovative production processes or the use of alternative energy sources. Low-impact materials minimize environmental harm during their life cycle, including resource extraction, manufacturing, use, and disposal. Examples include low-carbon concrete, low-VOC (volatile organic compounds) paints, and low-embodied energy insulation.

Non-Toxic and Health-Focused Materials: Non-toxic materials prioritize occupant health and well-being by minimizing the use of harmful chemicals and emissions. These materials are free from substances such as formaldehyde, lead, asbestos, and other hazardous compounds. Examples include zero-VOC adhesives, natural sealants, and non-toxic insulation materials.

Benefits of Green Building Materials

Green building materials offer numerous benefits that contribute to sustainable construction practices:

Energy Efficiency: Many green building materials have superior energy performance characteristics, contributing to energy-efficient buildings. For example, high-performance insulation reduces heat transfer and improves thermal comfort, while energy-efficient windows minimize heat loss or gain. These materials help reduce energy consumption for heating, cooling, and lighting, resulting in lower energy bills and reduced greenhouse gas emissions.

Resource Conservation: Green building materials prioritize resource efficiency and reduce the demand for virgin resources. Utilizing recycled or upcycled materials helps divert waste from landfills and decreases the need for resource extraction. Additionally, renewable materials support sustainable forestry practices and contribute to the preservation of ecosystems.

Improved Indoor Air Quality: Non-toxic materials and low-VOC products contribute to improved indoor air quality, creating healthier living and working environments. By minimizing emissions of harmful chemicals, green building materials help reduce the risk of respiratory issues, allergies, and other health concerns associated with poor indoor air quality.

Durability and Longevity: Green building materials often exhibit superior durability and longevity, reducing the need for frequent replacements and renovations. This not only reduces maintenance costs but also minimizes waste generation and conserves resources over the life cycle of a building.

Market Value and Branding: Incorporating green building materials can enhance the market value and marketability of a building. Increasingly, homebuyers, tenants, and investors are seeking environmentally responsible properties. Green buildings constructed with sustainable materials can command higher prices, attract environmentally conscious occupants, and strengthen the reputation of building owners and developers.

Green Building Technologies

In addition to green building materials, the integration of innovative technologies further enhances the sustainability and performance of buildings. These technologies leverage advanced systems, automation, and data analytics to optimize energy efficiency, improve occupant comfort, and reduce environmental impacts. Here are a few examples:

Smart Building Systems: Smart building systems utilize sensors, data analytics, and automation to optimize energy consumption, lighting, HVAC (heating, ventilation, and air conditioning), and overall building operations. These systems enable real-time monitoring and control, allowing adjustments based on occupancy, weather conditions, and energy demand.

Renewable Energy Systems: Integrating renewable energy technologies, such as solar panels, wind turbines, and geothermal systems, reduces reliance on fossil fuels and decreases greenhouse gas emissions. These systems generate clean, renewable energy on-site, contributing to energy independence and cost savings over time.

Water-Efficient Systems: Water-efficient technologies, including high-efficiency fixtures, rainwater harvesting systems, and graywater recycling systems, help reduce water consumption and promote sustainable water management. These systems optimize water use, minimize waste, and support water conservation efforts.

Building Energy Management Systems (BEMS): BEMS provide centralized control and monitoring of energy consumption, allowing building operators to identify energy-saving opportunities, track performance, and optimize building systems. These systems integrate various components, such as lighting controls, HVAC systems, and renewable energy sources, to maximize energy efficiency and reduce operational costs.

Green Roof and Living Wall Systems: Green roof and living wall systems involve the installation of vegetation on roofs or vertical surfaces, providing insulation, reducing stormwater runoff, enhancing air quality, and promoting biodiversity. These systems contribute to energy efficiency, improve microclimate, and create aesthetically pleasing and sustainable urban environments.

Design Considerations

When incorporating green building materials and technologies, several design considerations should be taken into account:

Whole-Building Approach: Sustainable design should consider the integration of green materials and technologies throughout the entire building, encompassing its envelope, interior finishes, systems, and landscaping. A holistic, whole-building approach ensures the maximum benefits in terms of energy efficiency, resource conservation, and occupant well-being.

Life Cycle Assessment: Conducting life cycle assessments of materials and technologies helps quantify their environmental impact and evaluate their sustainability credentials. Assessments consider factors such as resource consumption, energy use, emissions, and end-of-life disposal. Designers can utilize this information to make informed decisions and select the most sustainable options.

Adaptability and Future-Proofing: Designing for adaptability and future-proofing enables buildings to accommodate changing needs and technologies over time. Consider incorporating flexible layouts, modular systems, and adaptable infrastructure to support future upgrades or renovations without significant disruption or waste.

Passive Design Strategies: Integrating passive design strategies, such as proper orientation, shading, natural ventilation, and daylighting, enhances the energy efficiency and comfort of buildings. By utilizing passive design principles in conjunction with green building materials and technologies, designers can optimize building performance and reduce reliance on mechanical systems.

Education and Awareness: Educating building occupants about the benefits of green building materials and technologies fosters a culture of sustainability and encourages responsible behaviors. Providing information, guidelines, and training on energy-efficient practices and sustainable living helps occupants maximize the benefits of green buildings.

Green building materials and technologies are fundamental to sustainable construction practices. By prioritizing environmentally friendly materials, designers can reduce the ecological footprint of buildings, enhance energy efficiency, improve indoor air quality, and promote occupant health and well-being. Integrating innovative technologies further enhances building performance, allowing for optimalenergy management, water conservation, and occupant comfort. Through careful material selection, designers contribute to resource conservation, waste reduction, and the overall sustainability of the built environment. By utilizing green building materials and technologies, the construction industry can drive positive change, create healthier spaces, and pave the way for a more sustainable future.

CHAPTER 3: ENERGY EFFICIENCY AND RENEWABLE ENERGY INTEGRATION

3.1 Energy-Efficient Building Envelope Design

E nergy-efficient building envelope design is a crucial aspect of sustainable construction practices. The building envelope, consisting of exterior walls, roofs, windows, and doors, plays a significant role in regulating heat transfer, minimizing energy consumption, and enhancing occupant comfort. By adopting energy-efficient design principles and utilizing appropriate materials, designers can create buildings that optimize thermal performance, reduce reliance on mechanical heating and cooling systems, and contribute to overall energy efficiency. In this section, we will delve into the details of energy-efficient building envelope design, examining key considerations, strategies, and benefits.

Importance of Energy-Efficient Building Envelope

The building envelope acts as a barrier between the interior and exterior environments, and its design significantly impacts energy consumption and thermal comfort within a building. An energy-efficient building envelope offers several advantages:

Reduced Energy Consumption: A well-designed and properly insulated building envelope minimizes heat transfer through conduction, convection, and radiation. This reduces the reliance on mechanical heating and cooling systems, leading to lower energy consumption and reduced utility costs.

Enhanced Thermal Comfort: An energy-efficient building envelope helps maintain stable indoor temperatures, reducing temperature fluctuations and the need for frequent adjustments to heating and cooling systems. This contributes to improved thermal comfort for occupants, promoting productivity, well-being, and satisfaction.

Improved Indoor Air Quality: An effective building envelope design includes proper ventilation strategies, such as controlled air exchange and filtration, to ensure adequate indoor air quality. By minimizing air leakage and infiltration, the building envelope helps prevent the entry of outdoor pollutants and allergens, creating a healthier indoor environment.

Sustainability and Environmental Impact: Energy-efficient building envelopes align with sustainability goals by reducing greenhouse gas emissions associated with heating and cooling. By minimizing energy consumption, buildings with efficient envelopes contribute to mitigating climate change and reducing the overall environmental impact of the built environment.

Key Considerations in Building Envelope Design

When designing an energy-efficient building envelope, several key considerations should be taken into account:

Insulation: Proper insulation is critical for minimizing heat transfer and maximizing energy efficiency. The selection of insulation materials and their installation should be based on factors such as climate, building type, and desired thermal performance. Insulation can be added to walls, roofs, and floors to create a thermal barrier and reduce heat loss or gain.

Air Leakage Control: Controlling air leakage is essential for maintaining energy efficiency and indoor comfort. Air sealing measures, such as the use of weatherstripping, caulking, and sealing gaps around windows, doors, and penetrations, prevent the infiltration of outdoor air and the escape of conditioned air, reducing energy waste.

Windows and Glazing: Windows play a crucial role in the energy performance of a building envelope. Energy-efficient windows and glazing systems utilize advanced technologies, such as low-emissivity (low-E) coatings, multiple glazing layers, and gas fills (such as argon or krypton), to minimize heat transfer while maximizing natural light transmission. Proper window placement, shading devices, and orientation also contribute to energy efficiency.

Solar Heat Gain Control: Solar heat gain, the heat entering a building from sunlight, can significantly impact cooling loads. Strategies such as the use of shading devices, reflective coatings, and solar control glazing help reduce solar heat gain and prevent overheating, particularly in warm climates or buildings with extensive glazing.

Thermal Mass: Incorporating thermal mass, such as concrete or masonry, into the building envelope can help moderate temperature fluctuations by absorbing and releasing heat slowly. Thermal mass is effective in climates with significant diurnal temperature variations, contributing to more stable indoor temperatures and reduced reliance on mechanical heating and cooling.

Vapor Control: Proper vapor control measures, such as vapor barriers or vapor retarders, are essential to prevent moisture accumulation within the building envelope. By managing vapor diffusion, designers can protect insulation and structural components, mitigate the risk of condensation, and maintain the long-term integrity of the envelope.

Strategies for Energy-Efficient Building Envelope Design

To optimize energy efficiency and thermal performance, several strategies can be implemented in building envelope design:

Continuous Insulation: Providing continuous insulation across the entire building envelope minimizes thermal bridging, which occurs when there are breaks in insulation that allow heat transfer through structural elements. Continuous insulation, such as rigid foam insulation, ensures a continuous thermal barrier, reducing energy loss and improving overall efficiency.

High-Performance Glazing: Selecting energy-efficient windows and glazing systems with low U-factors and high solar heat gain coefficients (SHGC) can significantly improve the thermal performance of the building envelope. Considerations such as double or triple glazing, low-E coatings, and gas fills help reduce heat transfer and maximize energy savings.

Air Sealing: Ensuring airtightness in the building envelope is crucial for reducing energy waste due to air leakage. Proper air sealing measures, including sealing gaps, joints, and penetrations, help maintain indoor comfort, improve energy efficiency, and prevent moisture-related issues.

Passive Solar Design: Passive solar design principles, such as proper building orientation, strategic placement of windows, and shading devices, optimize solar heat gain and natural daylighting while minimizing heat loss or gain. Passive solar strategies can significantly reduce energy demand for heating and cooling, particularly in climates with ample sunshine.

Ventilation Strategies: Implementing efficient ventilation strategies, such as mechanical ventilation systems with heat recovery, helps maintain indoor air quality while minimizing energy loss. Heat recovery ventilation systems capture and transfer heat from exhaust air to incoming fresh air, reducing the energy required for heating or cooling.

Benefits of Energy-Efficient Building Envelope Design

Energy-efficient building envelope design offers numerous benefits that contribute to sustainable construction and building performance:

Energy Savings: A well-designed and properly insulated building envelope reduces heat loss in winter and heat gain in summer, resulting in reduced energy consumption for heating and cooling. This leads to significant energy savings and lower utility bills for building owners and occupants.

Comfort and Productivity: An energy-efficient building envelope helps maintain more stable indoor temperatures, minimizing hot or cold spots and temperature fluctuations. This improves occupant comfort, creating a conducive environment for productivity, concentration, and overall well-being.

Indoor Air Quality: By controlling air leakage and incorporating efficient ventilation systems, energy-efficient building envelopes help maintain a healthy indoor environment. Adequate ventilation ensures proper air exchange and filtration, reducing the presence of indoor pollutants and allergens.

Durability and Longevity: Properly designed and constructed energy-efficient building envelopes tend to be more durable and less prone to moisture-related issues, such as mold growth or deterioration. This enhances the longevity of the building, reducing the need for costly repairs or premature replacements.

Environmental Impact: Energy-efficient building envelopes contribute to reducing greenhouse gas emissions associated with heating and cooling, thereby mitigating climate change. By decreasing reliance on fossil fuel-based energy sources, buildings with efficient envelopes play a vital role in creating a more sustainable and low-carbon future.

Market Value and Branding: Energy-efficient buildings with high-performance envelopes have a competitive advantage in the real estate market. Green building certifications, such as LEED (Leadership in Energy and Environmental Design) or BREEAM (Building Research Establishment Environmental Assessment Method), recognize the significance of energy-efficient design and reward buildings with improved marketability and higher value.

Energy-efficient building envelope design is an integral part of sustainable construction practices. By prioritizing insulation, air sealing, window selection, solar heat gain control, and other design strategies, designers can create buildings that optimize thermal performance, reduce energy consumption, and enhance occupant comfort. The benefits of energy-efficient building envelopes include energy savings, improved indoor air quality, enhanced durability, and a reduced environmental impact. By incorporating these design principles, the construction industry can contribute to a more sustainable and energy-efficient built environment.

3.2 HVAC Systems and Energy Optimization

HVAC (Heating, Ventilation, and Air Conditioning) systems play a vital role in maintaining indoor comfort and air quality in buildings. However, they also consume a significant amount of energy, contributing to greenhouse gas emissions and energy costs. To promote sustainability and energy efficiency, it is crucial to optimize HVAC systems through innovative design, equipment selection, controls, and maintenance practices. In this section, we will explore the key considerations and strategies for optimizing HVAC systems to achieve energy efficiency, occupant comfort, and environmental sustainability.

Importance of HVAC System Optimization

HVAC systems account for a substantial portion of a building's energy consumption. By optimizing these systems, designers can reduce energy usage, decrease carbon emissions, and improve the overall performance of the building. Here are the key reasons why HVAC system optimization is important:

Energy Efficiency: Optimizing HVAC systems results in reduced energy consumption and lower operating costs. Energy-efficient equipment, intelligent controls, and effective maintenance practices contribute to improved energy performance, ensuring that HVAC systems operate at their highest efficiency levels.

Occupant Comfort: Well-designed and properly maintained HVAC systems provide optimal indoor thermal conditions, humidity control, and ventilation rates, enhancing occupant comfort and well-being. Maintaining comfortable indoor environments can improve productivity, health, and satisfaction among building occupants.

Indoor Air Quality: HVAC systems play a crucial role in maintaining good indoor air quality by controlling ventilation rates, filtering air pollutants, and managing humidity levels. Properly designed systems with effective filtration and ventilation strategies contribute to healthier indoor environments, reducing the risk of airborne pollutants and respiratory issues.

Environmental Impact: Energy consumption by HVAC systems contributes to greenhouse gas emissions and environmental degradation. Optimizing HVAC systems helps reduce carbon footprints, mitigate climate change, and conserve natural resources by minimizing energy waste and improving system efficiency.

Key Considerations for HVAC System Optimization

To optimize HVAC systems for energy efficiency and performance, the following key considerations should be taken into account:

Equipment Selection: Choosing energy-efficient HVAC equipment is critical to achieving optimal performance. Consider factors such as equipment efficiency ratings (e.g., SEER for cooling, AFUE for heating), energy-saving features, and compatibility with other system components. Energy-efficient equipment, such as variable-speed compressors, high-efficiency boilers, and heat recovery systems, can significantly reduce energy consumption.

Load Calculation and System Sizing: Proper load calculations are essential to determine the heating and cooling requirements of a building accurately. Oversized or undersized HVAC systems lead to inefficiencies, comfort issues, and increased energy consumption. Conducting load calculations considering factors like building orientation, insulation levels, occupancy, and equipment heat gains and losses ensures that HVAC systems are appropriately sized for optimal performance.

Zoning and Controls: Zoning and intelligent controls enable customized temperature and ventilation control in different areas of a building based on occupancy, usage patterns, and comfort requirements. By dividing the building into zones and providing independent controls, energy waste can be minimized, and occupant comfort can be optimized. Implementing advanced control strategies, such as occupancy sensors, setback schedules, and adaptive control algorithms, further enhances energy savings.

Ventilation Strategies: Proper ventilation is crucial for maintaining indoor air quality and occupant health. Implementing energy-efficient ventilation strategies, such as demand-controlled ventilation (DCV) or heat recovery ventilation (HRV), ensures that fresh air is provided when and where needed while minimizing energy waste associated with excessive ventilation rates.

Ductwork Design and Insulation: Well-designed and properly sealed ductwork minimizes energy losses due to air leakage and heat transfer. Optimizing duct layout, sizing, and insulation reduces pressure drops, improves airflow distribution, and minimizes thermal losses or gains. Sealing ducts effectively and insulating them properly contributes to energy efficiency and system performance.

Maintenance and Commissioning: Regular maintenance and commissioning of HVAC systems are essential to ensure optimal performance and energy efficiency throughout their lifecycle. Regular filter replacement, cleaning coils, lubricating components, and adjusting controls help maintain equipment efficiency and prevent energy waste. Commissioning ensures that the HVAC system is installed, calibrated, and operated according to design specifications.

Strategies for HVAC System

Optimization

To achieve optimal energy efficiency and performance in HVAC systems, the following strategies can be implemented:

Energy Recovery: Energy recovery systems, such as heat exchangers or heat pumps, capture and transfer waste heat from exhaust air to incoming fresh air or water. This reduces the energy required for heating or cooling and improves overall system efficiency.

Variable Speed Drives: Variable speed drives (VSDs) in HVAC equipment, such as fans and pumps, allow them to operate at variable speeds based on load demand. VSDs adjust motor speeds to match the required output, resulting in energy savings by avoiding constant full-speed operation.

Demand Response: Participating in demand response programs allows HVAC systems to respond to utility signals and adjust their operation during peak demand periods. By reducing energy consumption during peak hours, building owners can benefit from financial incentives while contributing to overall grid stability.

Advanced Controls and Building Automation Systems

Implementing advanced control systems and building automation systems (BAS) enables real-time monitoring, control, and optimization of HVAC system operation. These systems utilize sensors, algorithms, and predictive modeling to optimize energy consumption, occupancy comfort, and system performance.

Remote Monitoring and Analytics: Remote monitoring and analytics platforms provide real-time data on HVAC system performance, energy consumption, and equipment health. This allows for proactive maintenance, early detection of issues, and continuous optimization based on performance data analysis.

Benefits of HVAC System Optimization

Optimizing HVAC systems offers numerous benefits for building owners, occupants, and the environment:

Energy Savings: HVAC system optimization leads to reduced energy consumption, resulting in lower utility bills and long-term cost savings for building owners and occupants.

Improved Comfort: Well-designed and properly maintained HVAC systems ensure optimal indoor thermal conditions, humidity control, and air quality, enhancing occupant comfort and satisfaction.

Environmental Sustainability: Energy-efficient HVAC systems reduce greenhouse gas emissions, help mitigate climate change, and contribute to a more sustainable built environment.

Enhanced Indoor Air Quality: Proper ventilation and filtration strategies in optimized HVAC systems contribute to improved indoor air quality, reducing the presence of pollutants and allergens and promoting occupant health.

Long-Term Performance: HVAC system optimization improves equipment longevity, reduces the risk of breakdowns, and minimizes the need for costly repairs or replacements.

HVAC system optimization is essential for achieving energy efficiency, occupant comfort, and environmental sustainability in buildings. By considering equipment selection, load calculations, zoning, controls, ventilation strategies, ductwork design, and maintenance practices, designers can create HVAC systems that minimize energy consumption, improve indoor air quality, and enhance occupant comfort. Implementing advanced strategies, such as energy recovery, variable speed drives, demand response, and advanced controls, further maximizes energy savings and system performance. Optimized HVAC systems contribute to a sustainable built environment, reducing carbon footprints and ensuring long-term operational efficiency.

3.3 Integrating Renewable Energy Systems in Construction

The integration of renewable energy systems in construction plays a pivotal role in promoting sustainability, reducing carbon emissions, and achieving energy independence. Renewable energy technologies harness natural resources, such as sunlight, wind, and geothermal heat, to generate clean and renewable power. By incorporating these systems into the built environment, designers can create energy-efficient buildings that contribute to a low-carbon future. In this section, we will delve into the details of integrating renewable energy systems in construction, examining key technologies, considerations, and benefits.

Importance of Renewable Energy Integration

Integrating renewable energy systems in construction offers several important advantages:

Clean Energy Generation: Renewable energy systems harness abundant and naturally replenishing resources, such as sunlight and wind, to generate electricity or heat without relying on finite fossil fuels. By utilizing these clean energy sources, the environmental impact and carbon emissions associated with energy consumption are significantly reduced.

Energy Independence: Incorporating renewable energy systems in buildings promotes energy independence by generating on-site power. This reduces reliance on traditional energy grids and fossil fuel-based energy sources, enhancing the resilience and self-sufficiency of buildings, particularly in remote or off-grid locations.

Carbon Emission Reduction: Renewable energy systems produce electricity or heat with minimal or zero greenhouse gas emissions, mitigating the adverse effects of climate change. By displacing fossil fuel-based energy generation, buildings with integrated renewable systems contribute to global efforts to reduce carbon footprints and combat climate change.

Cost Savings: While the upfront costs of renewable energy systems can be higher than conventional energy sources, long-term cost savings can be achieved through reduced energy bills and potential incentives or rebates. Over time, the investment in renewable energy systems can yield financial benefits, offsetting or even surpassing the initial costs.

Key Technologies for Renewable Energy Integration

Several renewable energy technologies can be integrated into buildings to generate clean and sustainable power:

Solar Photovoltaic (PV) Systems: Solar PV systems convert sunlight into electricity through the use of solar panels composed of photovoltaic cells. These systems can be installed on roofs, facades, or as standalone structures. Solar PV systems can supply electricity for building operations, power appliances, and even feed surplus energy back into the grid through net metering arrangements.

Solar Thermal Systems: Solar thermal systems utilize sunlight to generate heat, which can be used for space heating, water heating, or industrial processes. These systems typically consist of solar collectors that absorb solar radiation and transfer heat to a fluid or air, which is then used for various purposes.

Wind Turbines: Wind turbines capture kinetic energy from wind and convert it into electricity. Small-scale wind turbines can be integrated into buildings or installed nearby to generate renewable power. Wind energy is particularly suitable for buildings in windy locations or open areas with favorable wind conditions.

Geothermal Systems: Geothermal systems tap into the natural heat beneath the Earth's surface to provide heating, cooling, and hot water. Ground-source heat pumps extract heat from the ground during winter and release heat into the ground during summer, leveraging the relatively stable temperature of the earth for efficient energy transfer.

Biomass Energy: Biomass energy utilizes organic matter, such as wood, agricultural residues, or dedicated energy crops, to produce heat, electricity, or biofuels. Biomass boilers or combined heat and power (CHP) systems can provide heating and electricity for buildings, utilizing renewable and sustainable fuel sources.

Hydropower Systems: Hydropower systems harness the energy of flowing or falling water to generate electricity. While large-scale hydropower is typically used in external power generation, smaller-scale hydro systems, such as micro-hydro turbines, can be integrated into buildings with access to a water source, such as rivers or streams.

Considerations for Renewable Energy Integration

When integrating renewable energy systems into construction projects, several key considerations should be taken into account:

Site Analysis: Conducting a site analysis helps identify the renewable energy potential, considering factors such as solar exposure, wind patterns, and available space. Understanding site-specific conditions allows for the selection of the most suitable renewable energy technologies and optimal system configurations.

Building Design and Orientation: The design and orientation of buildings can impact the effectiveness of renewable energy integration. Orienting buildings to maximize solar exposure or considering wind patterns can optimize the energy generation potential of solar PV or wind systems.

Load Analysis and Energy Efficiency: Conducting a thorough load analysis helps determine the energy demand of the building. Prioritizing energy efficiency measures, such as insulation, efficient lighting, and energy-saving appliances, reduces the overall energy demand and allows for the right-sized renewable energy system.

Grid Connection and Net Metering: For buildings connected to the grid, understanding the regulations and requirements for grid connection and net metering is essential. Net metering allows excess energy generated by renewable systems to be fed back into the grid, offsetting the energy consumed during times of low generation.

Financial and Regulatory Considerations: Assessing the financial viability and potential incentives for renewable energy integration is crucial. Understanding local regulations, incentives, and available financing options can facilitate the successful implementation of renewable energy systems.

Benefits of Renewable Energy Integration

Integrating renewable energy systems in construction offers numerous benefits:

Carbon Emission Reduction: Renewable energy systems reduce reliance on fossil fuels, leading to a significant reduction in carbon emissions and greenhouse gas footprints. By generating clean energy, buildings with integrated renewables contribute to mitigating climate change and transitioning to a low-carbon future.

Energy Independence and Resilience: On-site renewable energy generation enhances the energy independence and resilience of buildings. By producing their own power, buildings are less reliant on external energy grids, minimizing the impact of power outages and price fluctuations.

Cost Savings and Return on Investment: Renewable energy systems can provide long-term cost savings through reduced energy bills and potential incentives or rebates. Over time, the investment in renewable energy systems can pay off and deliver a return on investment.

Enhanced Building Value and Marketability: Buildings with integrated renewable energy systems often have higher market value and marketability. Sustainable features, including renewable energy, attract environmentally conscious buyers, tenants, and investors, enhancing the reputation and desirability of the building.

Environmental Stewardship and Corporate Social Responsibility: Integrating renewable energy systems demonstrates a commitment to environmental stewardship and corporate social responsibility. By utilizing clean and sustainable energy sources, building owners and developers contribute to a more sustainable future and set an example for others in the industry.

Integrating renewable energy systems in construction is crucial for achieving sustainability, reducing carbon emissions, and achieving energy independence. By harnessing the power of solar, wind, geothermal, biomass, or hydropower, buildings can generate clean and renewable energy on-site, reducing reliance on fossil fuels and contributing to a low-carbon future. Key considerations such as site analysis, building design, load analysis, and regulatory compliance play a significant role in successful integration. The benefits of renewable energy integration include carbon emission reduction, energy independence, cost savings, enhanced building value, and environmental stewardship. By embracing renewable energy technologies, the construction industry can lead the way toward a more sustainable and resilient built environment.

PART II: SUSTAINABLE CONSTRUCTION PRACTICES

CHAPTER 4: GREEN BUILDING CERTIFICATIONS AND STANDARDS

4.1 Leadership in Energy and Environmental Design

L eadership in Energy and Environmental Design (LEED) is a globally recognized green building certification program that promotes sustainable and environmentally responsible design, construction, and operation of buildings. Developed by the United States Green Building Council (USGBC), LEED provides a framework for assessing and rating the performance of buildings in various sustainability categories. In this section, we will delve into the details of LEED, examining its key principles, certification levels, assessment criteria, and benefits.

The Principles of LEED

LEED is built on a set of key principles that guide the design, construction, and operation of sustainable buildings. These principles encompass various aspects of sustainability and aim to minimize the environmental impact of the built environment:

Sustainable Site Development: LEED encourages site selection and development practices that minimize environmental disruption, preserve ecosystems, and promote sustainable land use.

Water Efficiency: LEED emphasizes the efficient use of water through strategies such as water-efficient landscaping, rainwater harvesting, and the use of low-flow fixtures and fittings.

Energy and Atmosphere: LEED promotes energy efficiency, the use of renewable energy sources, and the reduction of greenhouse gas emissions in building design and operation.

Materials and Resources: LEED encourages the use of environmentally friendly materials, waste reduction, recycling, and use of locally sourced materials to minimize the environmental impact of construction and demolition.

Indoor Environmental Quality: LEED focuses on creating healthy indoor environments through measures such as improved ventilation, the use of low-emitting materials, and enhanced occupant comfort and well-being.

Innovation: LEED encourages innovation in sustainable building practices, encouraging project teams to implement new technologies, strategies, and approaches that push the boundaries of environmental performance.

LEED Certification Levels

LEED offers different certification levels based on the overall sustainability performance of a building project. The certification levels provide a clear indication of a building's environmental performance and its commitment to sustainability:

LEED Certified: The entry-level certification level, LEED Certified, signifies that a building has met the basic requirements of sustainability and demonstrates a commitment to green building practices.

LEED Silver: The LEED Silver certification recognizes buildings that have achieved a higher level of sustainability, demonstrating a significant commitment to energy efficiency, resource conservation, and environmental responsibility.

LEED Gold: The LEED Gold certification represents an even higher level of sustainability achievement, showcasing exceptional performance in areas such as energy efficiency, water conservation, indoor air quality, and materials selection.

LEED Platinum: The highest certification level, LEED Platinum, is awarded to buildings that have demonstrated exceptional leadership in sustainability and have achieved the highest levels of performance across all LEED categories.

LEED Assessment Criteria

LEED certification is based on a comprehensive assessment of a building's sustainability performance. The assessment criteria cover a wide range of sustainability categories, evaluating various aspects of building design, construction, and operation. Some of the key assessment categories include:

Sustainable Sites: This category evaluates site selection, site development, stormwater management, and access to public transportation.

Water Efficiency: This category assesses water use reduction strategies, efficient fixtures and fittings, and innovative water management practices.

Energy and Atmosphere: This category focuses on energy

performance, renewable energy use, optimized HVAC systems, and greenhouse gas emissions reduction.

Materials and Resources: This category evaluates the use of sustainable materials, waste reduction strategies, recycling practices, and construction waste management.

Indoor Environmental Quality: This category assesses indoor air quality, daylighting, thermal comfort, low-emitting materials, and occupant comfort and satisfaction.

Innovation: This category recognizes innovative approaches, technologies, and strategies that go beyond the standard LEED requirements and contribute to sustainable building practices.

Benefits of LEED Certification

LEED certification offers numerous benefits to building owners, occupants, and the environment:

Environmental Performance: LEED-certified buildings have been shown to have lower energy consumption, reduced water usage, and decreased greenhouse gas emissions compared to conventional buildings. By adhering to LEED standards, buildings contribute to a healthier and more sustainable environment.

Energy Efficiency and Cost Savings: LEED certification promotes energy-efficient design, resulting in reduced energy consumption and lower operating costs. Building owners can benefit from lower utility bills and increased cost savings over the life of the building.

Improved Indoor Air Quality: LEED certification emphasizes the use of low-emitting materials, proper ventilation, and enhanced indoor air quality measures, creating healthier indoor environments for occupants. Improved indoor air quality contributes to occupant comfort, productivity, and well-being.

Marketability and Value: LEED-certified buildings have a competitive advantage in the real estate market. They attract environmentally conscious tenants and buyers, often commanding higher rental rates and sale prices. LEED certification demonstrates a commitment to sustainability, enhancing the marketability and value of the building.

Regulatory Recognition and Incentives: LEED certification is recognized by various governments, municipalities, and organizations around the world. Many jurisdictions offer incentives, grants, tax rebates, and expedited permitting processes for LEED-certified projects, encouraging the adoption of sustainable building practices.

Positive Public Image and Corporate Social Responsibility: LEED certification showcases an organization's commitment to environmental stewardship and corporate social responsibility. It enhances the organization's reputation, fosters positive public image, and aligns with sustainability goals.

LEED certification is a globally recognized program that promotes sustainable building practices and environmental responsibility. It provides a framework for assessing and rating the sustainability performance of buildings, considering various categories such as site development, water efficiency, energy, materials, indoor environmental quality, and innovation. LEED certification offers numerous benefits, including improved environmental performance, energy efficiency, cost savings, enhanced indoor air quality, marketability, regulatory recognition, and corporate social responsibility. By embracing LEED principles and pursuing certification, the construction industry can contribute to a more sustainable and resilient built environment.

4.2 Building Research Establishment Environmental Assessment Method

Building Research Establishment Environmental Assessment Method (BREEAM) is a widely recognized and respected environmental assessment methodology for buildings. Developed by the Building Research Establishment (BRE) in the United Kingdom, BREEAM evaluates the sustainability performance of buildings across a range of categories, providing a comprehensive framework for assessing, rating, and certifying their environmental credentials. In this section, we will delve into the details of BREEAM, exploring its key principles, assessment criteria, certification levels, and benefits.

The Principles of BREEAM

BREEAM is based on a set of key principles that guide the assessment and certification process. These principles encompass various aspects of sustainability and provide a robust framework for evaluating the environmental performance of buildings:

Holistic Approach: BREEAM takes a holistic approach to sustainability, considering environmental, social, and economic factors. It assesses the entire lifecycle of a building, from design and construction to operation and maintenance.

Adaptability: BREEAM recognizes that buildings should be adaptable to meet the changing needs of occupants and society over time. It encourages flexible design and construction approaches that enable future modifications and renovations.

Innovation: BREEAM promotes innovation in sustainable building practices, encouraging the integration of new technologies, materials, and strategies that push the boundaries of environmental performance.

Performance: BREEAM assesses the actual performance of buildings, considering energy consumption, water usage, waste generation, indoor air quality, and other environmental factors. It ensures that buildings meet or exceed minimum performance requirements.

Collaboration: BREEAM fosters collaboration among project teams, stakeholders, and industry professionals. It encourages engagement and cooperation to achieve sustainable design, construction, and operation of buildings.

BREEAM Assessment Categories

BREEAM assesses buildings across several categories to evaluate their sustainability performance. The assessment categories cover a wide range of environmental and social factors, ensuring a comprehensive evaluation of a building's overall sustainability. Some of the key assessment categories include:

Management: This category assesses the project's management processes, policies, and strategies, including sustainable procurement, stakeholder engagement, and environmental management systems.

Health and Well-being: This category evaluates factors that impact occupant health and well-being, such as indoor air quality, daylighting, thermal comfort, acoustic performance, and access to amenities and services.

Energy: The energy category focuses on energy efficiency, renewable energy use, and the reduction of greenhouse gas emissions. It considers factors such as building fabric performance, heating and cooling systems, lighting efficiency, and on-site renewable energy generation.

Water: This category assesses water consumption, water-efficient fixtures and fittings, rainwater harvesting, and the management of wastewater and stormwater.

Materials: The materials category evaluates the environmental impact of construction materials, considering factors such as embodied carbon, responsible sourcing, life cycle assessment, and waste reduction.

Land Use and Ecology: This category examines the impact of the building on the surrounding environment, including site selection, ecological protection, biodiversity enhancement, and sustainable landscaping.

Pollution: The pollution category addresses the prevention and mitigation of pollution, including measures to reduce air and water pollution, noise control, and the management of hazardous materials.

Transport: This category evaluates the accessibility and sustainability of transportation options for building occupants, considering factors such as public transportation access, bicycle facilities, electric vehicle charging infrastructure, and travel plans.

BREEAM Certification Levels

BREEAM offers different certification levels based on the overall sustainability performance of a building project. The certification levels provide a clear indication of a building's environmental performance and its commitment to sustainability:

Pass: The Pass level indicates that the project has achieved a minimum standard of environmental performance and meets the basic requirements of BREEAM.

Good: The Good level demonstrates that the project has achieved a higher level of sustainability, surpassing the minimum requirements and demonstrating a commitment to environmental best practices.

Very Good: The Very Good level represents a significant

achievement in sustainability, indicating that the project has performed exceptionally well across multiple assessment categories.

Excellent: The Excellent level signifies that the project has achieved an outstanding level of sustainability performance, exceeding the requirements of BREEAM and demonstrating industry-leading best practices.

Outstanding: The Outstanding level represents the highest level of sustainability achievement, showcasing exceptional performance across all BREEAM assessment categories. It signifies innovation, leadership, and a commitment to pushing the boundaries of environmental performance.

Benefits of BREEAM Certification

BREEAM certification offers numerous benefits to building owners, occupants, and the environment:

Environmental Performance: BREEAM-certified buildings are designed, constructed, and operated with a strong focus on sustainability. They have lower environmental impacts, reduced energy consumption, decreased water usage, and improved indoor environmental quality compared to conventional buildings.

Marketability and Branding: BREEAM certification enhances the marketability and branding of buildings. It demonstrates a commitment to sustainability, attracting environmentally conscious tenants, buyers, and investors who value sustainable design and operation.

Operational Cost Savings: BREEAM-certified buildings typically have lower operating costs due to reduced energy and water consumption. Energy-efficient design, renewable energy integration, and resource-efficient practices contribute to long-term cost savings for building owners.

Health and Well-being: BREEAM places a strong emphasis on occupant health and well-being. Certified buildings provide better indoor air quality, access to natural daylight, comfortable thermal conditions, and amenities that promote occupant well-being and satisfaction.

Regulatory Recognition and Incentives: BREEAM certification is widely recognized by governments, municipalities, and organizations around the world. Many jurisdictions offer incentives, grants, tax rebates, and expedited permitting processes for BREEAM-certified projects, encouraging the adoption of sustainable building practices.

Sustainable Construction Industry: BREEAM drives innovation and best practices in the construction industry. It fosters collaboration among project teams, encourages the use of sustainable materials and technologies, and promotes the adoption of environmentally responsible practices.

BREEAM is a comprehensive and highly respected environmental assessment method for buildings. It assesses buildings across multiple sustainability categories, promoting a holistic approach to sustainability. BREEAM certification provides a clear indication of a building's environmental performance and its commitment to sustainable design, construction, and operation. The benefits of BREEAM certification include improved environmental performance, marketability, operational cost savings, occupant health and well-being, regulatory recognition, and the promotion of sustainable practices in the construction industry. By embracing BREEAM principles and pursuing certification, the construction industry can contribute to a more sustainable and resilient built environment.

4.3 WELL Building Standard and Healthy Buildings

The WELL Building Standard is a performance-based system for measuring, certifying, and promoting features of the built environment that contribute to the health and well-being of occupants. Developed by the International WELL Building Institute (IWBI), the WELL Standard focuses on creating healthy indoor spaces that support physical, mental, and emotional well-being. In this section, we will delve into the details of the WELL Building Standard, exploring its key principles, certification process, categories, and the benefits of creating healthy buildings.

The Principles of the WELL Building Standard

The WELL Building Standard is founded on a set of principles that prioritize the health and well-being of building occupants. These principles guide the design, construction, and operation of healthy buildings:

Air: The air principle focuses on indoor air quality, emphasizing the reduction of contaminants, provision of fresh air, and effective ventilation systems to enhance occupant health and well-being.

Water: The water principle promotes access to clean and safe drinking water, as well as the efficient use and management of water resources within buildings to support health and hydration.

Nourishment: The nourishment principle encourages healthy eating habits by providing access to nutritious food options, promoting healthy food policies, and fostering environments that support mindful eating.

Light: The light principle emphasizes the use of natural and artificial lighting that supports the body's circadian rhythm, enhances visual comfort, and promotes alertness and productivity.

Fitness: The fitness principle encourages physical activity and movement by providing spaces, amenities, and programs that promote exercise, active transportation, and an active lifestyle.

Comfort: The comfort principle focuses on creating indoor environments that promote thermal, acoustic, and ergonomic comfort to enhance occupant satisfaction, productivity, and well-being.

Mind: The mind principle supports mental health and well-being by promoting spaces that enhance cognitive function, reduce stress, and provide opportunities for relaxation, privacy, and focus.

WELL Certification Process

The process of achieving WELL certification involves several steps that ensure the building meets the rigorous standards set by the WELL Building Standard. These steps typically include the following:

Registration: The first step is to register the project with the IWBI and establish the project team responsible for implementing the necessary measures to meet the standard.

Documentation: The project team gathers and submits documentation related to design, construction, and operations that demonstrate compliance with the requirements of the WELL Standard.

Performance Testing: Performance testing may be conducted to verify that the building meets the specified criteria. This can involve measurements of air quality, lighting levels, acoustic performance, and other relevant parameters.

Certification: Once the documentation and performance testing requirements are met, the project undergoes a review process by the IWBI. If the project meets the necessary criteria, it can receive WELL certification.

Categories of the WELL Building Standard

The WELL Building Standard addresses various aspects of the built environment that impact occupant health and well-being. These categories encompass different elements of a building's design, construction, and operation:

Air: The air category focuses on air quality, including factors such as ventilation, filtration, and control of pollutants to ensure a healthy indoor environment.

Water: The water category emphasizes access to clean and safe drinking water, as well as the implementation of water management strategies that promote conservation and healthy water systems.

Nourishment: The nourishment category encourages healthy eating habits through the provision of nutritious food options, access to clean and safe food preparation areas, and the promotion of healthy food policies.

Light: The light category addresses lighting design and optimization to support the body's natural circadian rhythm, enhance visual comfort, and promote alertness and well-being.

Fitness: The fitness category focuses on the promotion of physical activity and movement through the provision of fitness spaces, active design features, and wellness programs that encourage an active lifestyle.

Comfort: The comfort category encompasses elements such as thermal comfort, acoustic performance, and ergonomic design to create indoor environments that support occupant comfort and well-being.

Mind: The mind category addresses mental health and well-being, promoting spaces that enhance cognitive function, reduce stress, and provide opportunities for relaxation and restoration.

Benefits of Creating Healthy Buildings

Creating healthy buildings that align with the principles of the WELL Building Standard offers numerous benefits for occupants, building owners, and the environment:

Occupant Health and Well-being: Healthy buildings contribute to the physical, mental, and emotional well-being of occupants. By providing high-quality indoor environments, these buildings can enhance productivity, reduce stress, and support overall health.

Increased Productivity and Performance: Well-designed and healthy buildings have been shown to improve occupant productivity, cognitive function, and overall performance. Access to natural light, good indoor air quality, and comfortable environments can positively impact concentration, focus, and creativity.

Employee Satisfaction and Retention: Providing a healthy and supportive work environment can lead to increased employee satisfaction and retention. Healthy buildings demonstrate an organization's commitment to the well-being of its employees, fostering a positive and inclusive culture.

Energy Efficiency and Cost Savings: Many strategies that contribute to a healthy building, such as efficient lighting systems, optimized HVAC systems, and natural ventilation, also promote energy efficiency. Energy-efficient buildings not only reduce environmental impact but also result in long-term cost savings for building owners.

Environmental Sustainability: Healthy buildings often incorporate sustainable design and construction practices, reducing resource consumption and environmental impact. By integrating energy-efficient systems, water conservation measures, and sustainable materials, these buildings contribute to a more sustainable future.

Positive Public Image and Branding: Creating healthy buildings demonstrates a commitment to occupant health, sustainability, and social responsibility. Such buildings can enhance the public image and brand reputation of organizations, attracting environmentally conscious tenants, buyers, and investors.

The WELL Building Standard provides a comprehensive framework for creating healthy buildings that prioritize occupant health, well-being, and productivity. By incorporating principles such as air quality, water management, nourishment, light, fitness, comfort, and mind, the WELL Standard ensures that buildings support the physical, mental, and emotional needs of occupants. The benefits of creating healthy buildings include improved occupant health, increased productivity, employee satisfaction, energy efficiency, environmental sustainability, and a positive public image. By embracing the principles of the WELL Building Standard, the construction industry can play a significant role in promoting healthier and more sustainable built environments.

CHAPTER 5: WATER MANAGEMENT AND CONSERVATION

5.1 Rainwater Harvesting and Graywater Systems

Rainwater harvesting and graywater systems are sustainable water management practices that contribute to water conservation and reduce reliance on freshwater resources. These systems capture and utilize rainwater and graywater for non-potable uses, such as irrigation, toilet flushing, and outdoor cleaning. In this section, we will delve into the details of rainwater harvesting and graywater systems, exploring their benefits, components, design considerations, and implementation.

Rainwater Harvesting

Rainwater harvesting is the practice of collecting and storing rainwater for future use. It is a simple and effective way to reduce dependence on traditional water sources and promote water conservation. The process typically involves the following components:

Collection Surfaces: Rainwater is collected from various surfaces, such as rooftops, paved areas, and landscaped areas. These surfaces act as catchment areas, directing rainwater into the harvesting system.

Gutters and Downspouts: Gutters and downspouts are installed on rooftops to collect rainwater and channel it towards the storage system. They help ensure efficient collection and prevent water loss.

Filtration: Rainwater undergoes filtration to remove debris, leaves, and other impurities before it enters the storage system. Filtration systems can include mesh screens, sediment filters, and first-flush devices.

Storage Tanks: Rainwater is stored in tanks or cisterns for later use. These tanks can be above ground, underground, or built into the structure. The size of the storage system depends on the anticipated water demand and the available space.

Distribution System: A distribution system is used to deliver the harvested rainwater to its intended non-potable uses. This system may include pumps, pipes, and control valves to regulate the flow of water.

Graywater Systems

Graywater refers to wastewater generated from sources other than toilets, such as sinks, showers, and laundry. Graywater systems capture and treat this water for reuse, reducing the demand for fresh water and minimizing the strain on wastewater treatment facilities. The components of a graywater system typically include:

Collection: Graywater is collected from various sources within the building, such as bathroom sinks, showers, and washing machines. Separate plumbing lines are installed to divert the graywater to the treatment and storage system.

Treatment and Filtration: Graywater undergoes treatment and filtration processes to remove contaminants and pathogens. Common treatment methods include physical filtration, biological treatment, and disinfection.

Storage: Treated graywater is stored in dedicated storage tanks or cisterns. The size of the storage system depends on the anticipated graywater generation and the intended non-potable uses.

Distribution: A distribution system delivers the treated graywater to its designated non-potable uses, such as toilet flushing, irrigation, or industrial processes. Separate plumbing lines and fixtures are used to ensure that graywater is not mixed with the potable water supply.

Benefits of Rainwater Harvesting and Graywater Systems

Rainwater harvesting and graywater systems offer numerous benefits for both individuals and the environment:

Water Conservation: These systems reduce the demand for freshwater resources by utilizing alternative water sources for non-potable uses. By capturing and reusing rainwater and graywater, significant amounts of potable water can be conserved.

Cost Savings: Rainwater harvesting and graywater systems can lead to cost savings by reducing the need for supplemental water sources, lowering water bills, and decreasing reliance on municipal water supplies.

Reduced Strain on Infrastructure: By decreasing the demand for potable water, rainwater harvesting and graywater systems reduce the strain on water supply infrastructure, including treatment plants, distribution networks, and wastewater treatment facilities.

Sustainable Landscaping: Utilizing harvested rainwater for irrigation purposes promotes sustainable landscaping practices. It helps maintain green spaces and gardens without depleting freshwater resources.

Environmental Benefits: Rainwater harvesting and graywater systems reduce the strain on rivers, lakes, and groundwater sources, preserving these valuable ecosystems. They also help prevent stormwater runoff, which can contribute to water pollution and flooding.

Drought Resilience: During periods of water scarcity or drought, rainwater harvesting and graywater systems provide an alternative water source, helping to ensure a reliable supply for non-potable uses.

Design Considerations and Implementation

Designing and implementing rainwater harvesting and graywater systems require careful consideration of various factors:

Water Demand: Assessing the water demand for non-potable uses is crucial in determining the size of the storage tanks and the overall system capacity.

Site and Climate: The location and climate of the building influence the amount of rainwater available for harvesting. Adequate catchment areas and storage capacity should be designed accordingly.

Water Quality: Proper filtration and treatment processes should be employed to ensure the quality of harvested rainwater and treated graywater meets the required standards for the intended non-potable uses.

Plumbing Design: Separate plumbing systems should be installed to segregate graywater from the potable water supply and prevent cross-contamination. Backflow prevention devices are necessary to maintain water quality and safety.

Maintenance and Monitoring: Regular maintenance, such as tank cleaning, filter replacement, and system inspections, is essential to ensure the efficiency and longevity of rainwater harvesting and graywater systems.

Regulatory Requirements: Compliance with local regulations, building codes, and health and safety guidelines is necessary when designing and implementing these systems. Permits and approvals may be required for their installation.

Rainwater harvesting and graywater systems provide sustainable solutions for water conservation and management. These systems capture and utilize rainwater and graywater for non-potable purposes, reducing the demand for freshwater resources and promoting environmental sustainability. The benefits of rainwater harvesting and graywater systems include water conservation, cost savings, reduced strain on infrastructure, sustainable landscaping, environmental benefits, and drought resilience. When designing and implementing these systems, careful consideration of factors such as water demand, site conditions, water quality, plumbing design, and regulatory requirements is essential. By incorporating rainwater harvesting and graywater systems into buildings, individuals and communities can contribute to a more water-efficient and sustainable future.

5.2 Water-Efficient Plumbing Fixtures and Irrigation Practices

Water scarcity and the need for sustainable water management have led to increased focus on water-efficient plumbing fixtures and irrigation practices. Water-efficient fixtures reduce water consumption in buildings, while efficient irrigation practices optimize water use in landscapes and gardens. In this section, we will delve into the details of water-efficient plumbing fixtures and irrigation practices, exploring their benefits, types, design considerations, and implementation.

Water-Efficient Plumbing Fixtures

Water-efficient plumbing fixtures are designed to minimize water consumption without compromising functionality or user experience. These fixtures are commonly used in residential, commercial, and institutional buildings to reduce water waste. Some key types of water-efficient plumbing fixtures include:

Low-Flow Toilets: Low-flow toilets use less water per flush compared to traditional toilets. They employ innovative design features such as dual flush mechanisms, pressure-assisted flushing, or improved bowl and trapway designs to achieve water savings.

Water-Saving Faucets: Water-saving faucets incorporate aerators or flow restrictors to reduce water flow rates without affecting water pressure. These fixtures deliver a mix of air and water, maintaining the desired flow while conserving water.

Efficient Showerheads: Efficient showerheads restrict water flow while providing an enjoyable shower experience. They often feature flow restrictors or air infusion technologies that reduce water usage without compromising shower quality.

Waterless Urinals: Waterless urinals eliminate the need for flushing by using gravity and a sealant liquid or cartridge to control odors and prevent the release of urine odors into the restroom. They significantly reduce water consumption in commercial and public facilities.

Sensor-Operated Fixtures: Sensor-operated fixtures, such as sensor faucets and sensor-flush toilets, use occupancy sensors to activate water flow or flushing only when needed. This technology minimizes water waste resulting from faucets or toilets being left running.

Benefits of Water-Efficient Plumbing Fixtures

Water-efficient plumbing fixtures offer several benefits for both individuals and the environment:

Water Conservation: The primary benefit of water-efficient plumbing fixtures is the significant reduction in water consumption. By using less water, these fixtures contribute to water conservation, particularly in regions prone to water scarcity or with limited water resources.

Cost Savings: Water-efficient fixtures help reduce water bills, leading to cost savings for building owners and occupants. Over time, the reduced water consumption can result in substantial financial benefits, making them a wise investment.

Environmental Sustainability: Water-efficient fixtures contribute to environmental sustainability by reducing the strain on freshwater resources. By conserving water, these fixtures help protect ecosystems, maintain water quality, and preserve natural habitats.

Energy Savings: Reduced water consumption also translates into energy savings. Less water needs to be heated, resulting in lower energy requirements for water heating systems, such as boilers or water heaters.

Reduced Wastewater Generation: Water-efficient fixtures reduce the volume of wastewater generated, alleviating the burden on wastewater treatment facilities and reducing the environmental impact associated with wastewater treatment and discharge.

Irrigation Practices

Efficient irrigation practices play a crucial role in optimizing water use in landscapes, gardens, and agricultural settings. By employing appropriate irrigation methods and technologies, water can be used more efficiently to support plant growth while minimizing waste. Some key irrigation practices include:

Drip Irrigation: Drip irrigation delivers water directly to plant root zones, minimizing evaporation and runoff. It involves the use of tubing or drip emitters that slowly release water near the plant's base. This method reduces water waste and promotes healthier plant growth.

Micro-Sprinklers: Micro-sprinklers provide efficient water distribution by emitting water in fine droplets, reducing evaporation and wind drift. They are suitable for smaller areas or plants that require a broader coverage pattern.

Smart Irrigation Controllers: Smart irrigation controllers use weather data, soil moisture sensors, or evapotranspiration rates to adjust irrigation schedules and watering durations automatically. This technology ensures that plants receive the right amount of water based on actual needs, avoiding overwatering or underwatering.

Soil Moisture Sensors: Soil moisture sensors measure soil moisture content and help determine when irrigation is required. By providing real-time data on soil moisture levels, these sensors prevent excessive watering and promote water-efficient irrigation practices.

Mulching: Applying mulch to garden beds and landscaped areas helps retain soil moisture, reducing the frequency and amount of water needed for irrigation. Mulch also suppresses weed growth, preventing competition for water resources.

Design Considerations and Implementation

Designing and implementing water-efficient plumbing fixtures and irrigation practices require careful consideration of various factors:

Water Demand: Assessing the water demand for plumbing fixtures and irrigation systems is crucial in determining the appropriate fixture types, flow rates, and irrigation methods.

Regulatory Compliance: Compliance with local water efficiency regulations and building codes is necessary when selecting and installing water-efficient fixtures. Understanding and adhering to these regulations ensures compliance with water conservation guidelines.

Retrofitting and Upgrading: Retrofitting existing buildings with water-efficient fixtures involves assessing the compatibility and feasibility of retrofit options. Upgrading fixtures can lead to immediate water savings without the need for major plumbing modifications.

Maintenance and Education: Regular maintenance, including inspections and repairs, is essential to ensure that water-efficient fixtures and irrigation systems function optimally. Additionally, educating occupants and users about the benefits and proper use of these fixtures and practices can further enhance water conservation efforts.

Monitoring and Evaluation: Monitoring water usage and implementing water metering systems can help track consumption and identify areas for improvement. Regular evaluation of the effectiveness of water-efficient fixtures and irrigation practices allows for adjustments and optimization.

Water-efficient plumbing fixtures and irrigation practices play a crucial role in water conservation and sustainable water management. These practices reduce water consumption, promote cost savings, contribute to environmental sustainability, and optimize water use in buildings and landscapes. By incorporating water-efficient fixtures and implementing efficient irrigation methods, individuals, building owners, and communities can make significant contributions to water conservation efforts. Careful consideration of water demand, regulatory compliance, retrofitting options, maintenance, and monitoring is essential for the successful implementation of water-efficient plumbing fixtures and irrigation practices.

5.3 Stormwater Management Techniques

Stormwater management is a critical aspect of sustainable construction and urban planning. Effective stormwater management techniques aim to minimize the adverse impacts of stormwater runoff, reduce the strain on drainage systems, and protect water quality in nearby water bodies. In this section, we will explore various stormwater management techniques, including their benefits, types, design considerations, and implementation.

Importance of Stormwater Management

Stormwater runoff occurs when rainwater flows over land surfaces, roads, and rooftops, eventually finding its way into storm drains and water bodies. Uncontrolled stormwater runoff can lead to several issues, including:

Erosion and Sedimentation: Unmanaged stormwater runoff

can cause erosion of soil and sedimentation in rivers, streams, and other water bodies. This can impact aquatic ecosystems, degrade water quality, and impair habitats.

Flooding: Excessive stormwater runoff can overwhelm drainage systems, leading to localized flooding in urban areas. Flooding can damage infrastructure, disrupt daily activities, and pose risks to human safety.

Water Pollution: Stormwater runoff can pick up pollutants such as sediment, chemicals, heavy metals, and nutrients as it flows over impervious surfaces. If not managed properly, these pollutants can contaminate water bodies, affecting aquatic life and human health.

Streambank Erosion: Uncontrolled stormwater runoff can cause increased stream flow, leading to streambank erosion and destabilization of adjacent ecosystems.

Stormwater Management Techniques

Stormwater management techniques aim to address these challenges by effectively managing stormwater runoff and minimizing its adverse impacts. Some commonly used stormwater management techniques include:

Green Infrastructure: Green infrastructure refers to the use of vegetation, soil, and natural processes to manage stormwater. Techniques such as green roofs, rain gardens, bioswales, and permeable pavements help capture and absorb stormwater, allowing it to infiltrate into the ground and be naturally filtered.

Detention and Retention Ponds: Detention and retention ponds are engineered structures designed to temporarily store stormwater runoff. Detention ponds hold and slowly release stormwater to prevent overwhelming drainage systems, while retention ponds permanently retain a portion of stormwater, allowing for gradual infiltration and evaporation.

Infiltration Basins and Trenches: Infiltration basins and trenches are shallow depressions or excavated areas filled with permeable materials such as gravel or porous soil. They allow stormwater to infiltrate into the ground, promoting groundwater recharge and reducing runoff volumes.

Constructed Wetlands: Constructed wetlands mimic natural wetland ecosystems and help manage stormwater runoff. These wetlands utilize vegetation, soil, and microbes to remove pollutants and enhance water quality before the water is discharged into receiving water bodies.

Bioretention Cells: Bioretention cells, also known as rain gardens or bio-swales, are landscaped depressions filled with soil and vegetation. They capture and treat stormwater runoff, promoting infiltration and filtering out pollutants.

Permeable Pavements: Permeable pavements, such as permeable concrete, asphalt, or interlocking pavers, allow stormwater to infiltrate through the surface into underlying layers, reducing runoff and promoting groundwater recharge.

Benefits of Stormwater Management Techniques

Implementing effective stormwater management techniques offers several benefits, including:

Flood Mitigation: By managing stormwater runoff and reducing its volume and rate of flow, these techniques help prevent or mitigate localized flooding, protecting infrastructure and ensuring public safety.

Water Quality Improvement: Stormwater management techniques help remove pollutants and sediments from runoff before it reaches water bodies. This improves water quality, protects aquatic ecosystems, and enhances the overall health of water resources.

Groundwater Recharge: Techniques such as infiltration basins, permeable pavements, and green infrastructure promote groundwater recharge by allowing stormwater to infiltrate into the ground. This helps replenish aquifers and maintain the water table.

Aesthetic and Recreational Benefits: Green infrastructure elements, such as rain gardens and bioswales, enhance the aesthetic appeal of urban spaces. They also provide opportunities for recreational activities and contribute to the overall well-being of communities.

Reduced Drainage Infrastructure Costs: By effectively managing stormwater runoff at the source, these techniques help reduce the strain on drainage systems and the need for costly infrastructure expansion or upgrades.

Design Considerations and Implementation

Designing and implementing stormwater management techniques require careful consideration of various factors:

Site Assessment: A comprehensive assessment of the site, including topography, soil conditions, and hydrological characteristics, is essential to determine the appropriate stormwater management techniques.

Regulatory Compliance: Compliance with local stormwater management regulations and guidelines is necessary when designing and implementing stormwater management techniques. Understanding and adhering to these regulations ensures the project meets the required standards.

System Sizing: Proper sizing of stormwater management systems, such as ponds, basins, or infiltration areas, is crucial to ensure their effectiveness. Factors such as anticipated stormwater volumes, site constraints, and infiltration rates should be considered.

Maintenance and Monitoring: Regular maintenance and monitoring are essential to ensure the continued functionality and effectiveness of stormwater management techniques. This includes inspections, sediment and debris removal, vegetation maintenance, and monitoring of water quality parameters.

Integration with Site Design: Stormwater management techniques should be integrated into the overall site design and landscape planning. They should complement other site features, such as buildings, walkways, and recreational spaces, while providing functional and aesthetic benefits.

Stormwater management techniques play a crucial role in mitigating the adverse impacts of stormwater runoff and promoting sustainable water management. By implementing strategies such as green infrastructure, detention/retention ponds, infiltration systems, constructed wetlands, and permeable pavements, the negative effects of stormwater runoff can be minimized. These techniques offer benefits such as flood mitigation, water quality improvement, groundwater recharge, aesthetic enhancements, and reduced infrastructure costs. Careful consideration of site conditions, regulatory requirements, system sizing, maintenance, and integration with site design is necessary for successful implementation. By incorporating effective stormwater management techniques, communities can ensure the long-term sustainability of their water resources and create resilient urban environments.

CHAPTER 6: WASTE MANAGEMENT AND RECYCLING

6.1 Construction Waste Management Strategies

Construction waste is a significant environmental concern that arises from construction and demolition activities. Effective construction waste management strategies are essential to minimize the environmental impact, promote resource conservation, and enhance sustainability in the construction industry. In this section, we will explore various construction waste management strategies, including their benefits, types, best practices, and implementation.

Importance of Construction Waste Management

Construction activities generate a substantial amount of waste, including materials, debris, packaging, and discarded equipment. Improper handling and disposal of construction waste can have several negative impacts, including:

Environmental Pollution: Construction waste can release hazardous substances, chemicals, and pollutants into the environment, contaminating soil, water bodies, and air quality. This can harm ecosystems, wildlife, and human health.

Resource Depletion: Construction waste represents a loss of valuable resources. Proper waste management can help recover and reuse materials, reducing the need for new resource extraction and promoting resource conservation.

Landfill Burden: Improper disposal of construction waste can lead to excessive landfill usage and contribute to the depletion of landfill space. This increases the environmental and economic costs associated with waste management.

Energy Consumption: The production, transportation, and disposal of construction waste require significant energy inputs. By implementing waste management strategies, energy consumption associated with waste can be reduced.

Construction Waste Management Strategies

Construction waste management strategies aim to minimize waste generation, promote recycling and reuse, and ensure responsible disposal. Some commonly used construction waste management strategies include:

Waste Minimization and Source Separation: Waste minimization starts at the design stage by optimizing material usage, reducing excess packaging, and environmentally friendly products. Source separation involves segregating different types of waste on-site to facilitate recycling and proper disposal.

Material Recovery and Recycling: Material recovery and recycling involve salvaging and recycling construction waste materials, such as concrete, wood, metal, and plastics. These materials can be processed and reused in new construction projects, reducing the demand for virgin resources.

Donation and Reuse: Materials that are in good condition and no longer needed can be donated to local organizations, charities, or community projects. This promotes reuse, extends the lifespan of materials, and benefits the community.

Composting: Organic waste generated during construction, such as wood scraps and landscaping debris, can be composted. Composting converts organic waste into nutrient-rich compost that can be used for landscaping or soil improvement.

Waste-to-Energy Conversion: In cases where waste cannot be recycled or reused, waste-to-energy conversion technologies can be employed. These technologies convert waste into energy through processes such as incineration or anaerobic digestion, reducing reliance on fossil fuels.

Hazardous Waste Management: Proper handling, storage, and disposal of hazardous materials, such as asbestos, lead-based paint, and chemical solvents, are critical to ensure worker and environmental safety. Hazardous waste should be managed in accordance with local regulations and guidelines.

Best Practices in Construction Waste Management

Implementing effective construction waste management requires adherence to best practices. Some key best practices include:

Waste Management Plan: Develop a comprehensive waste management plan that outlines waste reduction goals, strategies for waste separation and recycling, responsible disposal procedures, and designated personnel responsible for waste management.

Training and Education: Provide training and educational programs for construction personnel to raise awareness about waste management practices, proper waste sorting, and the benefits of recycling and resource conservation.

On-Site Waste Sorting Stations: Set up designated waste sorting stations on construction sites to facilitate the separation of different types of waste, including recyclables, hazardous materials, and general waste. Clearly label and provide easy-to-use containers for each waste category.

Partner with Recycling Facilities: Establish partnerships with local recycling facilities or waste management companies to ensure the proper collection, transportation, and recycling of construction waste materials.

Regular Waste Audits: Conduct periodic waste audits to assess the effectiveness of waste management practices, identify areas for improvement, and track progress towards waste reduction targets.

Collaboration and Communication: Foster collaboration among project stakeholders, including contractors, subcontractors, suppliers, and waste management service providers. Effective communication and coordination ensure that waste management practices are consistently implemented throughout the construction project.

Implementation Considerations

Successful implementation of construction waste management strategies requires careful consideration of various factors:

Regulatory Compliance: Familiarize yourself with local regulations, permits, and waste management guidelines to ensure compliance with legal requirements. Obtain necessary permits and approvals for waste handling and disposal activities.

Contractor and Supplier Engagement: Engage contractors, subcontractors, and suppliers in waste management practices. Incorporate waste management requirements into contracts and establish expectations for waste reduction, recycling, and responsible disposal.

Waste Tracking and Reporting: Implement a waste tracking and reporting system to monitor the quantity and types of waste generated, recycled, and disposed of. This information can help evaluate the effectiveness of waste management efforts and provide data for reporting purposes.

Continuous Improvement: Continuously evaluate and improve waste management practices based on lessons learned, emerging technologies, and industry best practices. Stay updated on advancements in recycling technologies and waste reduction strategies.

Community Engagement: Engage with local communities and stakeholders to create awareness about construction waste management practices and their environmental benefits. Encourage community involvement in recycling initiatives and collaborate with local recycling centers.

Effective construction waste management strategies are crucial for minimizing environmental impact, promoting resource conservation, and enhancing sustainability in the construction industry. By implementing waste minimization, material recovery and recycling, donation and reuse, composting, waste-to-energy conversion, and proper hazardous waste management, construction professionals can significantly reduce the environmental footprint of their projects. Adhering to best practices, such as developing a waste management plan, training personnel, establishing waste sorting stations, and conducting waste audits, ensures the successful implementation of waste management strategies.

Collaboration, communication, regulatory compliance, and continuous improvement are essential elements for achieving sustainable construction waste management. By adopting these strategies and practices, the construction industry can contribute to a more sustainable and environmentally responsible future.

6.2 Recycling and Reuse of Construction Materials

Recycling and reuse of construction materials are essential practices in sustainable construction. These practices aim to reduce waste generation, conserve natural resources, and minimize the environmental impact associated with construction and demolition activities. By diverting materials from landfills and incorporating them back into the construction process, recycling and reuse offer significant environmental and economic benefits. In this section, we will explore the importance, benefits, methods, and challenges of recycling and reusing construction materials.

Importance of Recycling and Reuse

Construction and demolition activities generate a substantial amount of waste, including concrete, wood, metals, plastics, and other materials. By recycling and reusing these materials, the construction industry can address the following challenges:

Waste Reduction: Recycling and reusing construction materials help divert waste from landfills, reducing the strain on waste disposal facilities and mitigating environmental pollution.

Resource Conservation: By recycling and reusing materials, the demand for virgin resources is reduced. This promotes resource conservation and minimizes the need for resource extraction, thus preserving natural habitats and ecosystems.

Energy Savings: Recycling and reusing construction materials require less energy compared to manufacturing new materials from scratch. This leads to significant energy savings and reduces greenhouse gas emissions associated with resource extraction and production processes.

Cost Savings: Recycling and reusing materials can lead to cost savings for construction projects. By avoiding the need to purchase new materials and paying for waste disposal, project costs can be reduced.

Benefits of Recycling and Reuse

Implementing recycling and reuse practices in construction projects offers several benefits:

Environmental Conservation: Recycling and reusing materials help conserve natural resources, reduce energy consumption, lower greenhouse gas emissions, and minimize the environmental impact associated with resource extraction and manufacturing processes.

Waste Diversion: Recycling and reusing construction materials divert waste from landfills, reducing the volume of waste that ends up in disposal facilities. This extends the lifespan of landfills and reduces the need for new landfill sites.

Resource Efficiency: By incorporating recycled and reused materials, the construction industry can optimize resource efficiency and reduce reliance on virgin materials. This helps conserve finite resources and promotes a circular economy.

Reduced Environmental Footprint: Recycling and reusing materials contribute to the reduction of greenhouse gas emissions, air and water pollution, and environmental degradation associated with traditional manufacturing and waste disposal processes.

Economic Opportunities: Recycling and reuse activities create economic opportunities, such as the establishment of recycling facilities, material recovery centers, and markets for recycled and reclaimed materials. This stimulates job creation and supports local economies.

Methods of Recycling and Reuse

Various methods and techniques are employed to recycle and reuse construction materials. Some commonly used methods include:

Material Separation and Sorting: Construction waste is sorted and separated into different material types, such as concrete, wood, metals, and plastics. This allows for efficient recycling and reuse of specific materials.

Crushing and Grinding: Concrete and masonry waste can be crushed and ground into aggregates, which can be used as a substitute for natural aggregates in new concrete or as a base material for road construction.

Wood Reclamation and Repurposing: Wood waste can be reclaimed, processed, and repurposed for various applications, such as furniture, flooring, paneling, and landscaping materials.

Metal Recycling: Metals, including steel, aluminum, and copper, can be recycled and used in the production of new metal products or as raw materials in other industries.

Plastic Recycling: Plastic waste generated during construction can be sorted, processed, and recycled into new plastic products, such as pipes, containers, or composite building materials.

Salvaging and Reuse: Salvaging materials from demolition sites or renovation projects allows for their reuse in other construction projects. This includes salvaging fixtures, doors, windows, and architectural elements.

Challenges and Considerations

Despite the benefits and potential of recycling and reusing construction materials, several challenges exist:

Quality and Contamination: Ensuring the quality and purity of recycled materials is crucial for their successful reuse. Contamination from hazardous substances, improper sorting, or mixed waste streams can hinder the recycling and reuse process.

Economic Viability: The economic viability of recycling and reusing materials depends on factors such as market demand, transportation costs, processing technologies, and the availability of local recycling facilities. These factors can influence the feasibility and cost-effectiveness of recycling and reuse practices.

Awareness and Education: Creating awareness among construction professionals, project owners, and the general public about the benefits and importance of recycling and reusing construction materials is crucial for promoting these practices. Education and training programs can help overcome barriers and encourage broader adoption.

Building Codes and Standards: Building codes and standards play a vital role in facilitating the use of recycled and reclaimed materials in construction projects. Updating and harmonizing codes to incorporate sustainability considerations and allow for the use of recycled materials is necessary.

Collaboration and Supply Chain Integration: Collaboration among stakeholders, including contractors,subcontractors, suppliers, waste management companies, and recycling facilities, is essential for the successful implementation of recycling and reuse practices. Integration of the supply chain ensures seamless coordination, efficient material recovery, and optimal utilization of recycled and reclaimed materials.

Implementation Considerations

To effectively implement recycling and reuse practices in construction projects, several considerations should be taken into account:

Waste Management Plan: Develop a comprehensive waste management plan that includes strategies for recycling and reuse. This plan should outline the goals, responsibilities, procedures, and timelines for implementing recycling and reuse practices.

Material Identification and Sorting: Properly identify and sort construction waste to ensure the efficient recovery of recyclable and reusable materials. Provide designated areas for material separation and establish protocols for waste sorting on construction sites.

Collaboration and Partnerships: Establish partnerships with local recycling facilities, waste management companies, and suppliers of recycled materials. Collaborate with these stakeholders to ensure the effective recycling and reuse of construction materials.

Material Testing and Certification: Verify the quality and suitability of recycled and reclaimed materials through testing and certification processes. This ensures that the materials meet the necessary standards and can be safely and effectively incorporated into construction projects.

Monitoring and Evaluation: Implement a system to track and measure the amount of construction waste recycled and the volume of materials reused. Regularly evaluate the effectiveness of recycling and reuse practices to identify areas for improvement and set targets for waste reduction.

Continuous Improvement: Stay updated on advancements in recycling technologies, material recovery processes, and best practices in the industry. Continuously assess and improve recycling and reuse strategies based on lessons learned and emerging innovations.

Recycling and reuse of construction materials are essential practices for sustainable construction. By diverting waste from landfills and incorporating recycled and reclaimed materials, the construction industry can reduce its environmental footprint, conserve resources, and promote a circular economy. The benefits of recycling and reuse include waste reduction, resource conservation, energy savings, reduced environmental impact, and economic opportunities. However, challenges such as quality control, economic viability, awareness, and collaboration need to be addressed to realize the full potential of recycling and reuse in the construction industry. By implementing effective strategies, engaging stakeholders, and promoting a culture of sustainability, the construction sector can contribute to a greener and more sustainable future.

Implementing Circular Economy Principles in Construction

The concept of a circular economy has gained significant traction in recent years as a sustainable approach to resource management. In the construction industry, embracing circular economy principles offers tremendous potential for reducing waste, conserving resources, and promoting a more sustainable and resilient built environment. This section will explore the importance, benefits, strategies, and challenges of implementing circular economy principles in construction.

The Importance of Circular Economy in Construction

The construction industry is known for its substantial resource consumption and waste generation. By embracing circular economy principles, construction practices can be transformed to operate within the boundaries of a regenerative and sustainable system. Circular economy in construction aims to shift from the traditional linear model of "take-make-dispose" to a circular model of "reduce-reuse-recycle." This shift is critical for the following reasons:

Resource Conservation: Construction projects consume vast amounts of materials, energy, and water. Embracing circular economy principles allows for the efficient and responsible use of resources, minimizing extraction, and preserving natural ecosystems.

Waste Reduction: Construction activities generate significant amounts of waste, including demolition debris, excess materials, and packaging. A circular economy approach emphasizes waste reduction, reuse, and recycling, diverting materials from landfills and reducing the environmental impact.

Carbon Footprint Reduction: The construction sector is a significant contributor to global greenhouse gas emissions. By adopting circular economy principles, emissions can be reduced by optimizing material use, promoting energy efficiency, and minimizing waste generation.

Economic Opportunities: Circular economy practices in construction create new economic opportunities through the development of recycling and remanufacturing industries, job creation, and the emergence of innovative business models.

Benefits of Implementing

Circular Economy Principles

The implementation of circular economy principles in construction offers numerous benefits, including:

Resource Efficiency: Circular economy principles prioritize resource efficiency through the use of recycled and reclaimed materials, the extension of product lifecycles, and the reduction of waste. This minimizes the demand for virgin resources, conserves energy, and mitigates environmental degradation.

Waste Diversion: Circular economy practices encourage the reuse, refurbishment, and recycling of construction materials, diverting them from landfills. This reduces the strain on waste management systems, conserves landfill space, and decreases associated environmental impacts.

Cost Savings: Circular economy principles can lead to cost savings for construction projects. By reusing materials, minimizing waste disposal costs, and optimizing resource use, project expenses can be reduced.

Innovation and Collaboration: The circular economy encourages collaboration and innovation among stakeholders in the construction industry. This fosters the development of new technologies, business models, and partnerships that drive sustainability and resilience in the built environment.

Enhanced Resilience: Circular economy practices promote a more resilient construction industry by reducing reliance on scarce resources, mitigating supply chain disruptions, and improving resource management in the face of environmental and economic uncertainties.

Strategies for Implementing Circular Economy Principles

To successfully implement circular economy principles in construction, several strategies can be adopted:

Design for Reuse and Adaptability: Emphasize design principles that enable easy disassembly, modular construction, and adaptability to future needs. This allows for the reuse of building components and materials in future projects, reducing waste and extending their lifecycle.

Material Recovery and Recycling: Establish systems for the efficient recovery and recycling of construction materials. This involves sorting and separating materials on-site, partnering with recycling facilities, and incorporating recycled content in new construction.

Collaborative Business Models: Foster collaboration among stakeholders, including designers, contractors, suppliers, and waste management companies. Collaborative business models, such as material banks and resource-sharing platforms, facilitate the exchange and reuse of construction materials and equipment.

Prefabrication and Modular Construction: Embrace off-site prefabrication and modular construction techniques. This approach promotes resource efficiency, reduces waste generation, and enables the easy disassembly and reuse of building components.

Product and Material Innovation: Encourage the development of innovative products and materials that are designed with circular economy principles in mind. This includes exploring alternative, renewable, and recyclable materials, as well as promoting the use of environmentally friendly construction products.

Challenges and Considerations

Despite the benefits, implementing circular economy principles in construction is not without challenges:

Cultural Shift: Shifting from a linear to a circular economy mindset requires a cultural shift within the construction industry. This includes raising awareness, changing traditional practices, and fostering a commitment to sustainability at all levels.

Collaboration and Coordination: Circular economy principles require collaboration and coordination among various stakeholders, including designers, contractors, suppliers, waste management companies, and regulators. Effective communication and partnership are essential for successful implementation.

Knowledge and Skills: Adopting circular economy practices may require acquiring new knowledge and skills. This includes understanding circular design principles, implementing waste management strategies, and staying informed about innovative materials and technologies.

Regulatory Framework: The development of supportive regulatory frameworks and policies is necessary to incentivize circular economy practices in the construction industry. This includes providing incentives, setting targets, and establishing standards that promote resource efficiency and waste reduction.

Infrastructure and Technology: Adequate infrastructure and technologies are needed to support the recycling, reuse, and remanufacturing of construction materials. Investments in recycling facilities, sorting technologies, and advanced material recovery systems are essential for effective implementation.

Implementation Considerations

To effectively implement circular economy principles in construction, the following considerations should be taken into account:

Education and Awareness: Educate construction professionals about circular economy concepts, benefits, and best practices. Encourage training and knowledge sharing to build capacity and foster a culture of sustainability within the industry.

Collaborative Approach: Encourage collaboration and partnerships among stakeholders to develop shared goals and initiatives. Foster an environment where knowledge, resources, and experiences are shared to drive collective action towards circular economy implementation.

Life Cycle Assessment: Conduct life cycle assessments to evaluate the environmental impacts of construction materials and processes. This helps identify areas for improvement, prioritize actions, and make informed decisions that support circularity.

Policy Support: Advocate for supportive policies and regulations that incentivize circular economy practices in the construction sector. Engage with policymakers, industry associations, and relevant stakeholders to drive policy changes that promote sustainability and resource efficiency.

Monitoring and Evaluation: Implement monitoring and evaluation systems to track progress, measure the effectiveness of circular economy initiatives, and identify areas for improvement. Regularly assess and report on key performance indicators related to waste reduction, resource conservation, and circularity.

Implementing circular economy principles in construction is crucial for promoting sustainability, resource efficiency, and waste reduction. By embracing practices such as design for reuse, material recovery and recycling, collaborative business models, and innovation, the construction industry can transition towards a more circular and sustainable future. While challenges exist, addressing them through education, collaboration, supportive policies, and investments in infrastructure and technology will pave the way for a resilient and environmentally responsible construction sector. By integrating circular economy principles into construction practices, we can build a more sustainable and regenerative built environment for future generations.

CHAPTER 7: GREEN SITE DEVELOPMENT AND LANDSCAPING

7.1 Sustainable Site Planning and Development

S ustainable site planning and development are integral components of environmentally responsible construction practices. Sustainable site planning and development refers to the process of designing, developing, and managing construction projects in a manner that minimizes negative environmental impacts, maximizes resource efficiency, and promotes the well-being of both the natural environment and the community. It involves considering various factors, such as site selection, design, construction practices, and long-term management, with the goal of creating a built environment that is sustainable, resilient, and harmonious with its surroundings.

Sustainable site planning and development focuses on integrating principles of environmental stewardship, resource conservation, and social responsibility into every stage of a construction project. It encompasses strategies that aim to reduce pollution, protect natural resources, preserve biodiversity, optimize energy and water efficiency, enhance human health and well-being, and foster community engagement. This section will delve into the importance of sustainable site planning and development, key considerations, strategies, and benefits associated with this approach.

Importance of Sustainable Site Planning and Development

Sustainable site planning and development play a crucial role in promoting environmental stewardship and creating healthier communities. The built environment has a significant impact on natural ecosystems, water resources, energy consumption, and human well-being. By adopting sustainable practices, construction projects can address these concerns and achieve the following:

Environmental Protection: Sustainable site planning and development help protect natural habitats, wildlife, and ecosystems by minimizing land disturbance, preserving open spaces, and reducing pollution and waste generation.

Resource Conservation: By considering the site's natural features and incorporating sustainable design principles, construction projects can optimize resource use, minimize water consumption, preserve biodiversity, and reduce energy demand.

Climate Resilience: Sustainable site planning takes into account climate considerations, such as extreme weather events, temperature fluctuations, and precipitation patterns. This promotes the resilience of the built environment to climate change impacts.

Improved Quality of Life: Sustainable site planning and development prioritize human well-being by creating healthy, accessible, and aesthetically pleasing spaces. This includes considerations such as pedestrian-friendly design, access to green spaces, and efficient transportation systems.

Key Considerations in Sustainable Site Planning and Development

Effective sustainable site planning and development require careful consideration of various factors:

Site Selection: Choose sites that minimize environmental impact and optimize natural resources. Consider factors such as proximity to existing infrastructure, transportation options, access to amenities, and potential impacts on sensitive ecosystems.

Site Analysis: Conduct a comprehensive analysis of the site's physical features, including topography, soil quality, vegetation, and water resources. Assess potential constraints and opportunities for sustainable design and development.

Water Management: Develop strategies for efficient water management, including rainwater harvesting, stormwater management, and water-efficient landscaping. Minimize runoff, reduce water consumption, and protect water quality through proper treatment and filtration.

Energy Efficiency: Integrate energy-efficient design principles into site planning and development. Consider factors such as solar orientation, shading, natural ventilation, and the use of renewable energy sources to reduce energy demand and mitigate climate impact.

Biodiversity and Habitat Preservation: Identify and protect significant natural features, habitats, and biodiversity on the site. Incorporate native vegetation, create wildlife corridors, and implement measures to minimize the impact on ecological systems.

Access and Mobility: Design sites that prioritize pedestrian and bicycle-friendly infrastructure, promote public transportation options, and provide convenient access to amenities and services. Encourage active transportation modes to reduce reliance on private vehicles and minimize carbon emissions.

Strategies for Sustainable Site Planning and Development

To achieve sustainable site planning and development, several strategies can be employed:

Site Design and Layout: Optimize site design to maximize open spaces, preserve existing vegetation, and promote connectivity. Consider factors such as building orientation, clustering, and the integration of green infrastructure elements.

Green Space and Landscaping: Incorporate green spaces, parks, and landscaped areas into site plans. Use native and drought-tolerant plant species, employ water-efficient irrigation systems, and create multifunctional landscapes that provide habitat, reduce heat island effects, and enhance aesthetics.

Stormwater Management: Implement sustainable stormwater management techniques, such as permeable pavement, bioswales, and rain gardens. These features help capture and filter stormwater runoff, recharge groundwater, and mitigate the risk of flooding and water pollution.

Material Selection and Construction Practices: Consider sustainable material choices that minimize environmental impact throughout the construction process. Prioritize locally sourced, recycled, or renewable materials and employ construction practices that reduce waste generation and pollution.

Site Rehabilitation and Restoration: Incorporate site rehabilitation and restoration measures to mitigate any negative impacts from construction activities. This includes restoring disturbed areas, replanting vegetation, and implementing erosion control measures.

Collaboration and Stakeholder Engagement: Foster collaboration among project stakeholders, including developers, designers, local communities, and regulatory agencies. Involve stakeholders in the site planning process, gather input, and incorporate community needs and aspirations into the development plans.

Benefits of Sustainable Site Planning and Development

Implementing sustainable site planning and development practices yield several benefits:

Environmental Preservation: By prioritizing environmental considerations, sustainable site planning and development help preserve natural resources, protect ecosystems, and conserve biodiversity.

Resource Efficiency: Sustainable site planning optimizes the use of resources, reduces water consumption, promotes energy efficiency, and minimizes waste generation. This leads to cost savings, reduced environmental impact, and enhanced long-term sustainability.

Enhanced Resilience: Sustainable site planning considers climate change impacts, such as extreme weather events and rising temperatures. By integrating resilience measures, projects are better equipped to withstand and adapt to future challenges.

Improved Livability: Sustainable sites promote healthier living environments, with features such as green spaces, pedestrian-friendly design, access to amenities, and reduced air and noise pollution. These factors contribute to improved quality of life for residents and users of the built environment.

Social and Economic Benefits: Sustainable site planning can stimulate economic development, create job opportunities, and enhance community cohesion. The integration of sustainable design and green infrastructure can attract investment, support local businesses, and improve the overall livability and attractiveness of an area.

Implementation Considerations

To effectively implement sustainable site planning and development, the following considerations should be taken into account:

Regulatory Compliance: Familiarize yourself with local planning regulations, zoning requirements, and environmental guidelines. Ensure that the proposed development aligns with applicable codes and standards.

Stakeholder Engagement: Engage with local communities, neighborhood groups, and relevant stakeholders throughout the planning and development process. Seek input, address concerns, and incorporate community aspirations into the design and decision-making.

Performance Measurement: Establish performance metrics and monitoring systems to track the sustainability outcomes of the site planning and development. Regularly assess and evaluate the performance of sustainable strategies and adjust as necessary.

Continual Learning and Improvement: Stay informed about emerging trends, innovative technologies, and best practices in sustainable site planning and development. Continually learn and adapt to new knowledge to improve project outcomes and promote sustainability.

Sustainable site planning and development are crucial for creating environmentally responsible, resource-efficient, and livable built environments. By considering key factors such as site selection, water management, energy efficiency, biodiversity preservation, and stakeholder engagement, construction projects can minimize environmental impact, conserve resources, and enhance community well-being. The integration of sustainable strategies offers numerous benefits, including environmental protection, cost savings, improved resilience, and social and economic advantages.

7.2 Green Roofs and Living Walls

Green roofs and living walls are innovative and sustainable design features that have gained popularity in recent years. These green infrastructure elements offer numerous environmental, social, and economic benefits, making them valuable additions to the built environment. In this section, we will explore green roofs and living walls in detail, discussing their definition, types, benefits, considerations for implementation, and their role in enhancing sustainability.

Definition and Types of Green Roofs and Living Walls

Green roofs, also known as vegetated roofs or rooftop gardens, are elevated surfaces covered with vegetation. They can be either intensive or extensive. Intensive green roofs are characterized by a deep soil layer that supports a diverse range of plant species, similar to a traditional garden. Extensive green roofs, on the other hand, have a shallower soil layer and typically support low-maintenance plants such as sedums, mosses, and grasses.

Living walls, also called green walls or vertical gardens, are vertical structures covered with plants that grow on a support system. They can be installed indoors or outdoors and come in various forms, including modular systems, panel systems, or trellis-based systems. Living walls can incorporate a variety of plant species, from small ferns and herbs to larger shrubs and vines.

Benefits of Green Roofs and Living Walls

Green roofs and living walls offer a range of benefits that contribute to the sustainability and well-being of the built environment:

Environmental Benefits

Improved Air Quality: Green roofs and living walls help purify the air by filtering pollutants and absorbing carbon dioxide (CO_2). Plants capture airborne particulate matter and release oxygen, creating a healthier and cleaner atmosphere.

Enhanced Stormwater Management: These green infrastructure elements can absorb and retain rainwater, reducing stormwater runoff and alleviating the burden on urban drainage systems. They help prevent flooding, decrease the strain on sewer systems, and promote groundwater recharge.

Urban Heat Island Mitigation: Green roofs and living walls reduce the urban heat island effect by providing shade, reducing surface temperatures, and releasing moisture through evapotranspiration. This contributes to cooler microclimates and improved thermal comfort in urban areas.

Biodiversity Promotion: These green features provide habitat and support biodiversity, attracting insects, birds, and other wildlife to urban environments. They help create ecological corridors and contribute to the conservation of native plant species.

Energy Efficiency

Thermal Insulation: Green roofs and living walls act as effective thermal insulators, reducing heat gain in summer and heat loss in winter. They improve building energy efficiency, leading to reduced heating and cooling needs and lower energy consumption.

Reduced HVAC Load: The presence of green roofs and living walls can decrease the demand for mechanical heating, ventilation, and air conditioning (HVAC) systems. This results in energy savings and lower greenhouse gas emissions.

Noise Reduction: These green features provide additional acoustic insulation, absorbing and attenuating noise from external sources. They help create quieter and more peaceful indoor and outdoor environments.

Health and Well-being

Improved Air Quality Indoors: Indoor green walls contribute to improved air quality by filtering and purifying indoor air, reducing volatile organic compounds (VOCs), and enhancing humidity levels. This leads to healthier and more pleasant indoor environments.

Biophilic Benefits: Green roofs and living walls provide visual and sensory connections with nature, promoting mental well-being, reducing stress, and enhancing productivity. They contribute to biophilic design principles that prioritize human connection with the natural environment.

Aesthetics and Property Value

Enhanced Aesthetics: Green roofs and living walls add visual appeal and beauty to buildings and urban landscapes. They can transform concrete jungles into vibrant and green spaces, improving the overall aesthetics of the built environment.

Increased Property Value: Properties with green roofs and living walls often experience increased value and market appeal. These green features are viewed as desirable and forward-thinking, attracting tenants, buyers, and investors.

Implementation Considerations for Green Roofs and Living Walls

Implementing green roofs and living walls requires careful planning, design, and consideration of various factors:

Structural Considerations: Assess the structural capacity of the building or structure to ensure it can support the additional weight of the green roof or living wall. Engage structural engineers and professionals to evaluate the feasibility and make necessary modifications if required.

Waterproofing and Drainage: Proper waterproofing and drainage systems are essential for the successful installation and long-term performance of green roofs and living walls. Ensure that appropriate waterproof membranes, root barriers, and drainage layers are in place to prevent water infiltration and damage.

Plant Selection and Maintenance: Select plant species that are suitable for the local climate, building orientation, and available sunlight. Consider factors such as plant hardiness, drought tolerance, and maintenance requirements. Regular maintenance, including watering, pruning, and monitoring for pests or diseases, is necessary to ensure the health and longevity of the green features.

Irrigation and Water Management: Implement efficient irrigation systems that supply water to green roofs and living walls based on plant needs. Consider rainwater harvesting and smart irrigation technologies to reduce water consumption and ensure optimal water management.

Professional Expertise: Engage experienced professionals, such as landscape architects, horticulturists, and green infrastructure specialists, to design and install green roofs and living walls. Their expertise is crucial in selecting appropriate plant species, designing irrigation systems, ensuring proper installation, and providing ongoing maintenance.

Code and Regulatory Compliance: Familiarize yourself with local building codes, regulations, and permits related to green roofs and living walls. Ensure compliance with fire safety, structural integrity, and environmental regulations.

Integration with Building Systems: Coordinate with architects, engineers, and contractors to integrate green roofs and living walls seamlessly with building systems, including waterproofing, HVAC, and electrical systems. Consider factors such as access for maintenance, compatibility with roof slopes, and potential impacts on rooftop equipment.

Role of Green Roofs and Living Walls in Enhancing Sustainability

Green roofs and living walls play a vital role in enhancing sustainability in the built environment:

Urban Greening: These green features contribute to urban greening efforts, addressing the challenges of urbanization, loss of green spaces, and environmental degradation. They help create more sustainable and livable cities by introducing vegetation and nature into densely populated areas.

Climate Change Adaptation and Mitigation: Green roofs and living walls help mitigate climate change by reducing energy consumption, mitigating the urban heat island effect, and sequestering carbon dioxide. They also enhance climate resilience by providing cooling effects, stormwater management, and supporting biodiversity.

Sustainable Urban Development: Green roofs and living walls align with principles of sustainable urban development, such as compact and efficient land use, mixed land-use planning, and the creation of healthy and inclusive environments. They contribute to the creation of sustainable, vibrant, and resilient communities.

Environmental Education: These green features offer opportunities for environmental education and awareness. Schools, universities, and public spaces with green roofs and living walls can serve as living laboratories, promoting an understanding of ecological processes, sustainable practices, and the importance of biodiversity.

Green Building Certifications: Green roofs and living walls contribute to meeting the criteria for various green building certifications, such as LEED (Leadership in Energy and Environmental Design) and BREEAM (Building Research Establishment Environmental Assessment Method). They enhance the sustainability performance of buildings and can contribute to achieving certification levels.

Green roofs and living walls are sustainable design elements that provide numerous environmental, social, and economic benefits. They improve air quality, manage stormwater, enhance energy efficiency, promote biodiversity, and contribute to human health and well-being. Implementing green roofs and living walls requires careful planning, expertise, and consideration of various factors. These green features play a significant role in enhancing sustainability in the built environment, contributing to urban greening, climate change mitigation and adaptation, and sustainable urban development. By using green roofs and living walls, we can create more sustainable, resilient, and harmonious cities that prioritize the well-being of both people and the planet.

7.3 Native Plants and

Ecosystem Restoration

Native plants and ecosystem restoration are vital components of sustainable land management and biodiversity conservation. By incorporating native plants into landscape design and implementing restoration practices, we can restore and enhance ecological systems, support native wildlife, and promote long-term sustainability. This section will explore the importance of native plants and ecosystem restoration, discuss the benefits they offer, and highlight key considerations for their successful implementation.

Importance of Native Plants in Ecosystem Restoration

Native plants are species that occur naturally in a particular region and have evolved in that specific ecological context. They play a critical role in ecosystem functioning and provide a range of benefits:

Biodiversity Conservation: Native plants are key components of local ecosystems, providing habitat, food sources, and nesting sites for native wildlife. By restoring native plant communities, we can support the biodiversity of an area and promote the survival of local species, including insects, birds, mammals, and pollinators.

Ecosystem Services: Native plants contribute to essential ecosystem services, such as carbon sequestration, soil stabilization, water filtration, and air purification. They have adapted to local conditions, making them well-suited for providing these services efficiently and effectively.

Adaptation to Climate and Environmental Conditions: Native plants have evolved to thrive in specific climate and soil conditions, making them more resilient to local climatic variations, drought, and pests. By utilizing native species, we can enhance the resilience of ecosystems and landscape designs in the face of climate change and other environmental stressors.

Soil Health and Nutrient Cycling: Native plants have complex root systems that help improve soil structure, increase organic matter content, and enhance nutrient cycling. They can contribute to soil fertility, reduce erosion, and improve water infiltration rates.

Cultural and Historical Significance: Native plants often hold cultural and historical significance for indigenous communities and local cultures. Their restoration and preservation help protect cultural heritage and maintain connections to traditional knowledge and practices.

Benefits of Ecosystem Restoration

Ecosystem restoration involves the intentional activity of assisting the recovery of an ecosystem that has been degraded, damaged, or destroyed. The restoration process aims to reestablish ecological functions, enhance biodiversity, and improve ecosystem resilience. Here are some key benefits of ecosystem restoration:

Habitat Restoration: Ecosystem restoration helps recreate and enhance habitat conditions for native wildlife. By reintroducing native plants, we can provide food, shelter, and nesting sites for a diverse range of species, including endangered and threatened ones.

Water Management: Restoring ecosystems can help regulate water flow, reduce erosion, and improve water quality. Native vegetation acts as a natural filter, reducing the amount of sediment, nutrients, and pollutants entering water bodies.

Carbon Sequestration: Ecosystem restoration can contribute to carbon sequestration, helping to mitigate climate change. Native plants have the capacity to store carbon in their biomass and in the soil, thus offsetting greenhouse gas emissions.

Enhanced Resilience: Restored ecosystems are often more resilient to disturbances, such as extreme weather events and invasive species. The presence of native plants increases the ecosystem's ability to recover from disturbances and adapt to changing conditions.

Aesthetics and Recreation: Restored ecosystems can provide aesthetically pleasing landscapes and recreational opportunities for communities. These areas offer spaces for nature appreciation, hiking, birdwatching, and other outdoor activities, promoting physical and mental well-being.

Considerations for Implementing Native Plant and Ecosystem Restoration

Successful implementation of native plant and ecosystem restoration requires careful planning, engagement with stakeholders, and consideration of site-specific factors:

Site Assessment: Conduct a thorough assessment of the site to understand its ecological characteristics, including soil composition, hydrological conditions, and existing vegetation. Identify any potential constraints or opportunities for restoration and prioritize areas for intervention.

Native Plant Selection: Choose native plant species that are well-adapted to the site's conditions, including soil type, moisture levels, sun exposure, and climate. Consider factors such as plant growth habits, life cycles, and their ability to provide habitat and food sources for native wildlife.

Propagation and Planting: Ensure a reliable supply of native plant material through proper propagation techniques, such as seed collection, nursery production, or sourcing from reputable suppliers. Follow best practices for planting, including appropriate planting techniques, soil preparation, and post-planting care to maximize plant survival rates.

Invasive Species Management: Identify and manage invasive plant species that could compete with or threaten the success of native plants. Implement strategies to control invasive species, such as manual removal, herbicide application, or biological control methods, to prevent their negative impacts on ecosystem restoration efforts.

Stakeholder Engagement: Engage with local communities, landowners, and stakeholders throughout the restoration process. Seek their input, share knowledge about the importance of native plants and ecosystem restoration, and foster a sense of ownership and stewardship.

Monitoring and Adaptive Management: Implement a monitoring program to assess the progress and success of ecosystem restoration efforts. Regularly monitor plant survival, biodiversity response, and ecosystem functioning indicators. Use this information to inform adaptive management strategies and adjust restoration practices as needed.

Restoration Techniques and Approaches

Several restoration techniques and approaches can be employed to achieve successful native plant and ecosystem restoration:

Direct Seeding: Directly seeding native plant species into the restoration site can be a cost-effective and efficient approach, particularly for larger areas. It involves distributing seeds across the site and ensuring proper seed-soil contact for germination and establishment.

Transplanting and Vegetation Propagation: Transplanting

established native plants or propagating them in a nursery setting can accelerate the restoration process. This approach ensures a higher survival rate of plants and can be particularly useful for areas with challenging site conditions or specific restoration goals.

Habitat Enhancement: Creating or enhancing specific habitat features, such as wetlands, meadows, or riparian areas, can be part of ecosystem restoration efforts. This involves modifying site conditions to mimic natural processes and create suitable conditions for native plants and wildlife.

Invasive Species Control: Effectively managing invasive species is crucial for successful ecosystem restoration. Implement a combination of techniques, such as manual removal, targeted herbicide application, or the introduction of biocontrol agents, to suppress invasive species and prevent their re-establishment.

Long-Term Maintenance and Management: Implement a long-term maintenance and management plan to ensure the continued success of ecosystem restoration efforts. This includes activities such as invasive species monitoring, prescribed burning, reseeding, and adaptive management based on ecological feedback.

Native plants and ecosystem restoration are fundamental elements of sustainable land management and biodiversity conservation. By restoring native plant communities, we can enhance ecological resilience, support native wildlife, and promote the long-term sustainability of our landscapes. Ecosystem restoration offers a range of benefits, including enhanced biodiversity, improved water management, carbon sequestration, and increased resilience to environmental changes. Successful implementation requires careful site assessment, stakeholder engagement, proper plant selection, and ongoing monitoring and adaptive management. By embracing native plants and ecosystem restoration, we can create healthier, more resilient, and sustainable ecosystems that support the well-being of both people and the natural world.

PART III: SUSTAINABLE CONSTRUCTION IN PRACTICE

CHAPTER 8: SUSTAINABLE PROJECT PLANNING AND MANAGEMENT

8.1 Incorporating Sustainability in Project Planning

Sustainability has become a critical consideration in the construction industry, driving the need for sustainable project planning. By integrating sustainability principles into the early stages of project planning, construction professionals can lay the foundation for environmentally responsible, socially conscious, and economically viable construction projects. This section will explore the importance of incorporating sustainability in project planning, discuss key strategies and considerations, and highlight the benefits it brings to the construction industry.

The Importance of Incorporating Sustainability in Project Planning

Incorporating sustainability in project planning is essential for several reasons:

Environmental Responsibility: Construction projects have significant environmental impacts, from resource consumption and energy use to waste generation and greenhouse gas emissions. By integrating sustainability in project planning, professionals can minimize these impacts, promote resource efficiency, and mitigate environmental degradation.

Social Consciousness: Sustainable project planning considers the social aspects of construction projects, aiming to create built environments that improve the quality of life for communities. It involves engaging stakeholders, promoting inclusivity, and prioritizing social well-being, health, and safety.

Economic Viability: Sustainable construction practices are not only environmentally and socially responsible but also economically beneficial. Incorporating sustainability in project planning can lead to cost savings, improved resource management, enhanced marketability, and long-term financial viability.

Regulatory Compliance and Certifications: Many jurisdictions now require or incentivize sustainable construction practices. By incorporating sustainability in project planning, construction professionals can ensure compliance with local regulations and take advantage of green building certifications such as LEED (Leadership in Energy and Environmental Design) and BREEAM (Building Research Establishment Environmental Assessment Method).

Key Strategies for Incorporating Sustainability in Project Planning

To effectively incorporate sustainability in project planning, construction professionals should consider the following strategies:

Set Clear Sustainability Goals: Establish specific and measurable sustainability goals for the project. These goals may include energy efficiency targets, waste reduction objectives, water conservation measures, and the use of sustainable materials. Clear goals provide a roadmap for decision-making throughout the project lifecycle.

Conduct a Sustainability Assessment: Perform a comprehensive assessment of the project's environmental, social, and economic aspects. This assessment may involve conducting an environmental impact assessment, evaluating social and community impacts, and analyzing life cycle costs. The results will help identify potential risks and opportunities for sustainable practices.

Engage Stakeholders: Engage stakeholders early in the project planning process to understand their needs, expectations, and concerns. Stakeholders may include clients, community members, regulatory agencies, design professionals, and construction teams. Collaboration and open communication foster a sense of ownership, build trust, and improve project outcomes.

Integrative Design Approach: Adopt an integrative design approach that considers the interdependencies and synergies among various project elements. This approach involves multidisciplinary collaboration and the integration of sustainable design principles from the outset. It ensures that sustainability considerations are embedded in the project's design, materials selection, and construction processes.

Life Cycle Thinking: Embrace life cycle thinking by considering the environmental, social, and economic impacts of a project from its inception to its end-of-life. Evaluate the life cycle impacts of various design and material choices, including energy use, carbon emissions, water consumption, waste generation, and maintenance requirements. Optimize choices to minimize life cycle impacts.

Sustainable Materials and Technologies: Give preference to sustainable materials and technologies that have a lower environmental footprint, such as recycled materials, renewable energy systems, energy-efficient equipment, and water-saving technologies. Consider life cycle assessment, durability, availability, and compatibility with the project's sustainability goals.

Energy and Resource Efficiency: Prioritize energy and resource efficiency measures throughout the project. This may involve optimizing building orientation, using passive design strategies to reduce energy demand, employing energy-efficient HVAC systems and lighting, and implementing water-efficient fixtures and irrigation systems. Energy modeling and analysis can help inform design decisions.

Waste Management and Recycling: Develop a waste management plan that emphasizes waste reduction, recycling, and responsible disposal. Set targets for waste diversion, promote on-site sorting and recycling, and explore opportunities for salvaging and reusing materials. Consider waste reduction strategies during construction, such as lean construction practices and prefabrication.

Green Procurement and Supply Chain Management

Give preference to suppliers and contractors with demonstrated sustainability commitments. Incorporate sustainable procurement practices by specifying environmentally friendly materials, promoting local sourcing, and considering the life cycle impacts of products and services.

Considerations for Sustainable

Project Planning

Incorporating sustainability in project planning requires careful consideration of several key factors:

Regulatory and Code Compliance: Familiarize yourself with local regulations, building codes, and green building certification requirements. Ensure that sustainability goals align with mandatory and voluntary standards. Seek guidance from sustainability experts or consultants to navigate complex requirements.

Cost-Benefit Analysis: Conduct a comprehensive cost-benefit analysis to evaluate the financial implications of sustainable measures. Consider both short-term and long-term costs and savings associated with energy efficiency, material selection, waste management, and other sustainable practices. Recognize that upfront investments in sustainability often yield significant long-term returns.

Knowledge and Training: Enhance the knowledge and skills of project stakeholders by providing training and education on sustainable construction practices. Engage with professional organizations, attend workshops or seminars, and encourage ongoing learning to stay up-to-date with emerging trends and best practices.

Monitoring and Evaluation: Implement a robust monitoring and evaluation plan to track progress towards sustainability goals. Collect data on energy consumption, water usage, waste generation, and other key performance indicators. Regularly review and analyze this data to identify areas for improvement and inform decision-making.

Continuous Improvement: Embrace a culture of continuous improvement by learning from past projects, sharing lessons learned, and implementing feedback mechanisms. Encourage innovation and explore emerging technologies, materials, and construction methods that offer even greater sustainability benefits.

Benefits of Incorporating Sustainability in Project Planning

Incorporating sustainability in project planning brings numerous benefits to the construction industry and society as a whole:

Environmental Stewardship: By integrating sustainable practices, construction projects contribute to environmental stewardship by reducing resource consumption, minimizing waste generation, and mitigating greenhouse gas emissions. This helps preserve natural resources and protect ecosystems.

Enhanced Reputation and Marketability: Projects that prioritize sustainability gain a competitive advantage in the marketplace. Clients, investors, and consumers increasingly value sustainable practices and may prefer projects that demonstrate environmental and social responsibility.

Improved Cost Efficiency: Sustainable project planning can lead to cost savings over the project's life cycle. Energy-efficient designs, reduced water consumption, waste management strategies, and long-term maintenance considerations can result in lower operational costs and increased financial viability.

Positive Social Impact: Sustainable projects prioritize the well-being and safety of the community. They create healthier, more livable spaces, promote social equity, and provide opportunities for local employment and skills development.

Resilient Infrastructure: Sustainable project planning incorporates resilience measures to anticipate and adapt to future challenges, such as climate change impacts and resource scarcity. By integrating resilience strategies, projects can withstand shocks and disturbances, ensuring long-term functionality and durability.

Incorporating sustainability in project planning is crucial for building sustainable construction practices. By considering environmental, social, and economic aspects from the outset, construction professionals can minimize environmental impacts, promote social well-being, and ensure long-term financial viability. Strategies such as setting clear sustainability goals, conducting assessments, engaging stakeholders, and adopting an integrative design approach can guide the process. Key considerations include regulatory compliance, cost-benefit analysis, knowledge and training, monitoring and evaluation, and a culture of continuous improvement. The benefits of incorporating sustainability in project planning are far-reaching, including environmental stewardship, enhanced reputation and marketability, improved cost efficiency, positive social impact, and resilient infrastructure. By prioritizing sustainability in project planning, the construction industry can make significant strides towards a more sustainable and resilient future.

8.2 Sustainable Procurement and Supply Chain Management

Sustainable procurement and supply chain management play a crucial role in promoting environmental and social responsibility within the construction industry. By adopting sustainable practices in procurement and supply chain processes, construction professionals can minimize negative impacts on the environment, support local economies, and enhance social well-being. This section will delve into the importance of sustainable procurement and supply chain management, explore key strategies and considerations, and highlight the benefits they bring to the construction industry.

The Importance of Sustainable Procurement and Supply Chain Management

Sustainable procurement and supply chain management are critical for several reasons:

Environmental Impact: Construction projects have a significant environmental footprint, largely influenced by the materials and resources used throughout the supply chain. By adopting sustainable procurement practices, such as selecting environmentally friendly materials, minimizing waste generation, and reducing carbon emissions, construction professionals can reduce the environmental impact of their projects.

Social Responsibility: Sustainable procurement and supply chain management prioritize the well-being of workers, local communities, and society as a whole. By ensuring fair labor practices, supporting local businesses, and engaging with suppliers committed to social responsibility, construction professionals can contribute to social equity and the overall welfare of stakeholders.

Legal Compliance: Sustainable procurement practices help construction companies meet legal obligations and adhere to regulations related to environmental protection, labor rights, and ethical sourcing. By proactively engaging in sustainable procurement, construction professionals can ensure compliance with local and international standards and avoid legal and reputational risks.

Market Demand and Reputation: Increasingly, clients, investors, and consumers expect sustainable practices from construction projects. By adopting sustainable procurement and supply chain management, construction companies can gain a competitive advantage, enhance their reputation, and attract environmentally and socially conscious clients.

Key Strategies for Sustainable Procurement and Supply Chain Management

To effectively integrate sustainability into procurement and supply chain management, construction professionals should consider the following strategies:

Define Sustainable Procurement Goals: Establish clear and measurable sustainability goals for procurement activities. These goals may include reducing carbon emissions, promoting the use of recycled or renewable materials, supporting local businesses, and ensuring fair and ethical sourcing practices. Clearly defined goals guide decision-making and enable the evaluation of performance.

Supplier Engagement and Evaluation: Engage suppliers early in the procurement process to assess their sustainability practices. Evaluate suppliers based on criteria such as environmental performance, social responsibility, ethical sourcing, and adherence to labor standards. Develop long-term relationships with suppliers committed to sustainability and encourage continuous improvement in their practices.

Sustainability Criteria in Bid Evaluation: Incorporate sustainability criteria into the bid evaluation process. Consider factors such as environmental impact, social responsibility, local content, life cycle costs, and innovation in supplier selection. Award contracts to suppliers who demonstrate a strong commitment to sustainability.

Material and Product Selection: Prioritize the use of sustainable materials and products that have a lower environmental impact. Consider factors such as resource efficiency, durability, recyclability, and toxicity. Give preference to materials with third-party certifications or eco-labels that validate their environmental and social attributes.

Local Sourcing and Community Engagement: Support local economies and communities by sourcing materials and services locally whenever feasible. This reduces transportation-related emissions, fosters economic development, and enhances community engagement. Collaborate with local suppliers, subcontractors, and workforce to create a positive social and economic impact.

Supply Chain Transparency: Promote transparency throughout the supply chain by encouraging suppliers to disclose information on their environmental and social performance. Seek visibility into the origin of materials, labor conditions, and the use of environmentally sensitive processes. Transparent supply chains enable the identification of potential risks, such as environmental violations or unethical practices.

Environmental Impact Assessment: Conduct environmental impact assessments for key materials and suppliers to identify potential environmental risks and evaluate their sustainability performance. Assess the carbon footprint, water usage, waste generation, and other environmental indicators associated with the supply chain. Use this information to inform decision-making and identify opportunities for improvement.

Collaboration and Partnerships: Collaborate with suppliers, industry associations, and other stakeholders to drive sustainable procurement practices collectively. Share best practices, collaborate on research and development initiatives, and explore joint sustainability projects. By working together, construction professionals can leverage collective knowledge and resources for sustainable supply chain management.

Continuous Improvement and Performance Measurement: Establish performance indicators and metrics to track the progress of sustainable procurement and supply chain management efforts. Regularly monitor and evaluate performance against established targets. Use the data collected to identify areas for improvement, set new goals, and implement strategies to drive continuous improvement.

Considerations for Sustainable Procurement and Supply Chain Management

When implementing sustainable procurement and supply chain management practices, construction professionals should consider the following factors:

Collaboration with Internal Stakeholders: Engage internal stakeholders, such as project managers, engineers, and procurement teams, in sustainable procurement efforts. Foster cross-functional collaboration to ensure alignment with project goals and to leverage expertise from different disciplines.

Capacity Building: Enhance the knowledge and skills of procurement teams and suppliers through training programs and capacity-building initiatives. Raise awareness about sustainable procurement practices, provide guidance on relevant standards and certifications, and promote the adoption of best practices throughout the supply chain.

Supplier Relationship Management: Develop strong relationships with suppliers based on trust, open communication, and shared sustainability goals. Regularly communicate expectations, provide feedback on performance, and collaborate on improvement strategies. Encourage suppliers to innovate and provide sustainable alternatives to existing products and materials.

Risk Management: Identify and manage potential risks associated with sustainable procurement, such as supplier non-compliance, environmental incidents, or labor violations.

Implement robust monitoring and auditing processes to ensure compliance with sustainability criteria andto identify any risks or non-conformities in the supply chain.

Verification and Certification: Consider third-party verification and certification schemes to validate the sustainability claims of suppliers. Certifications such as ISO 14001 (Environmental Management System) and SA8000 (Social Accountability) can provide assurance of environmental and social responsibility. Request documentation and evidence of certifications from suppliers as part of the procurement process.

Integration with Project Management Systems: Integrate sustainable procurement practices with project management systems to ensure seamless communication and coordination. Use project management tools and platforms to track sustainability-related information, monitor supplier performance, and facilitate collaboration between project teams and suppliers.

Benefits of Sustainable Procurement and Supply Chain Management

Implementing sustainable procurement and supply chain management practices brings numerous benefits to the construction industry and broader society:

Environmental Conservation: Sustainable procurement practices contribute to the conservation of natural resources, reduction of carbon emissions, and protection of ecosystems. By selecting environmentally friendly materials and minimizing waste generation, construction projects can help preserve the environment for future generations.

Social Responsibility: Sustainable procurement promotes fair labor practices, human rights, and social equity throughout the supply chain. By engaging with suppliers committed to social responsibility, construction professionals can support workers' rights, improve labor conditions, and foster inclusive and diverse workplaces.

Risk Mitigation: Sustainable procurement practices help mitigate risks associated with environmental incidents, supply chain disruptions, and reputational damage. By proactively assessing suppliers' sustainability performance, construction professionals can identify potential risks and take preventive measures to avoid disruptions and negative impacts.

Cost Savings and Efficiency: Sustainable procurement can lead to cost savings through improved resource efficiency, reduced waste, and optimized material selection. Energy-efficient products and materials can also lower operational costs over the lifecycle of a project. Sustainable practices drive efficiency and financial viability.

Stakeholder Satisfaction: Sustainable procurement practices align with the expectations of stakeholders, including clients, investors, employees, and local communities. By demonstrating a commitment to environmental and social responsibility, construction companies can enhance stakeholder satisfaction, build trust, and improve their reputation in the market.

Market Differentiation: Sustainable procurement and supply chain management can differentiate construction companies from their competitors. Clients increasingly prioritize sustainable practices and seek partners with demonstrated environmental and social responsibility. By positioning themselves as leaders in sustainability, construction professionals can gain a competitive edge and access new market opportunities.

Sustainable procurement and supply chain management are vital for driving environmental and social responsibility in the construction industry. By adopting sustainable practices, construction professionals can minimize environmental impacts, support local economies, and promote social well-being. Strategies such as setting clear sustainability goals, engaging suppliers, prioritizing sustainable materials, and fostering collaboration are essential for success. The benefits of sustainable procurement and supply chain management include environmental conservation, social responsibility, risk mitigation, cost savings, stakeholder satisfaction, and market differentiation. By embracing sustainable practices throughout the procurement and supply chain processes, the construction industry can make significant strides towards a more sustainable and responsible future.

8.3 Project Team Collaboration and Communication

Project team collaboration and communication are essential for the successful implementation of sustainable construction practices. In an industry where complex projects involve multiple stakeholders, effective collaboration and communication ensure that sustainability goals are understood, shared, and achieved. This section will delve into the importance of project team collaboration and communication in sustainable construction, explore key strategies and considerations, and highlight the benefits they bring to project outcomes.

The Importance of Project Team Collaboration and Communication

Project team collaboration and communication are crucial for several reasons:

Alignment of Sustainability Goals: Sustainable construction projects require a shared vision and understanding of sustainability goals among all project team members. Effective collaboration and communication ensure that everyone involved, from project managers to architects, engineers, contractors, and subcontractors, is aligned and working towards common sustainability objectives.

Integration of Diverse Expertise: Sustainable construction projects involve various disciplines and specialties, each contributing unique expertise. Effective collaboration allows for the integration of diverse perspectives, ensuring that sustainable design, construction, and operational strategies are considered holistically and optimized for the project's success.

Identification of Synergies and Trade-offs: Collaborative teamwork facilitates the identification of synergies and trade-offs between sustainability goals and project requirements. It enables discussions and decision-making processes that balance environmental, social, and economic considerations, leading to innovative and optimized solutions.

Conflict Resolution and Issue Management: Collaboration and open communication provide a platform for addressing conflicts and resolving issues promptly. By fostering a culture of trust, respect, and constructive dialogue, project teams can overcome challenges related to sustainability implementation and find mutually beneficial resolutions.

Stakeholder Engagement: Effective collaboration and communication extend beyond the project team to include engagement with external stakeholders such as clients, regulatory authorities, local communities, and NGOs. Involving stakeholders in sustainability discussions and decision-making processes fosters transparency, builds relationships, and ensures project outcomes align with broader societal needs and expectations.

Key Strategies for Project Team Collaboration and Communication

To foster effective collaboration and communication in sustainable construction projects, consider the following strategies:

Early Engagement and Integrated Design: Encourage early engagement of all relevant project team members, including architects, engineers, contractors, and sustainability consultants. Integrated design workshops and charrettes allow for collaborative brainstorming, idea-sharing, and the exploration of sustainable design strategies from the project's inception.

Clearly Defined Roles and Responsibilities: Clearly define roles and responsibilities of project team members, particularly regarding sustainability aspects. Ensure that each team member understands their contribution to sustainability goals and the specific tasks they are accountable for. This clarity helps prevent ambiguity and ensures efficient collaboration.

Effective Communication Channels: Establish effective communication channels to facilitate regular information sharing and updates. Utilize project management tools, collaborative platforms, and communication technologies to streamline communication and ensure timely dissemination of information across the project team.

Cross-Disciplinary Coordination: Promote cross-disciplinary coordination by facilitating regular meetings and workshops where team members from different disciplines can exchange knowledge, align strategies, and address sustainability-related challenges. Encourage collaboration and the sharing of best practices and lessons learned throughout the project lifecycle.

Open and Transparent Decision-Making Processes: Foster an environment of open and transparent decision-making, where all team members have the opportunity to contribute their expertise and opinions. Encourage constructive discussions, respect differing viewpoints, and seek consensus to ensure that sustainability decisions are informed, well-balanced, and aligned with project objectives.

Stakeholder Engagement and Consultation: Engage relevant stakeholders throughout the project, seeking their input, feedback, and involvement in sustainability-related discussions. Regularly communicate project progress, sustainability goals, and outcomes to stakeholders, inviting their perspectives and addressing their concerns. This engagement builds trust, improves decision-making, and enhances the project's social acceptance.

Continuous Learning and Training: Promote continuous learning and professional development among project team members, specifically in the area of sustainable construction practices. Provide training and workshops on sustainability concepts, emerging trends, and best practices to enhance team members' knowledge and skills, fostering innovation and a culture of sustainability.

Documentation and Knowledge Management: Establish effective documentation and knowledge management practices to capture and share project-related information, lessons learned, and best practices. This ensures that valuable insights and experiences are preserved, facilitating future projects and promoting continuous improvement in sustainable construction practices.

Considerations for Effective Collaboration and Communication: When implementing project team collaboration and communication strategies, consider the following factors:

Cultural and Language Diversity: Respect and accommodate cultural and language diversity within the project team. Foster an inclusive environment where team members feel comfortable expressing their thoughts and ideas, overcoming potential barriers that may arise due to cultural or language differences.

Clear Communication Protocols: Establish clear communication protocols that outline expectations, channels, and frequency of communication. Clearly define communication milestones, such as regular team meetings, progress reports, and sustainability updates, to ensure consistent and effective information exchange.

Conflict Resolution Mechanisms: Developclear conflict resolution mechanisms to address disagreements or conflicts that may arise during the project. Encourage open and respectful dialogue, establish a process for resolving conflicts, and involve relevant stakeholders when necessary to ensure fair and equitable solutions.

Flexibility and Adaptability: Recognize that project circumstances and sustainability requirements may evolve over time. Remain flexible and adaptable to changing needs, allowing for adjustments in collaboration and communication strategies as the project progresses. Embrace feedback and lessons learned to improve collaboration and address emerging challenges.

Leadership and Accountability: Strong leadership and clear accountability are essential for effective collaboration and communication. Project leaders should set the tone for collaborative behavior, establish a supportive and inclusive culture, and hold team members accountable for their sustainability-related responsibilities.

Benefits of Project Team Collaboration and Communication

Implementing effective collaboration and communication practices within the project team brings numerous benefits to sustainable construction projects:

Enhanced Sustainability Performance: Collaborative teamwork allows for the integration of diverse expertise and perspectives, leading to more robust sustainability solutions. By leveraging the collective knowledge and skills of the project team, sustainable construction goals can be effectively achieved.

Innovation and Creative Problem-Solving: Collaboration fosters innovation by encouraging the exploration of new ideas and approaches. Through collaborative problem-solving, project teams can find creative solutions to sustainability challenges, identifying opportunities for efficiency, resource optimization, and improved project outcomes.

Improved Project Efficiency: Effective communication and collaboration minimize misunderstandings, reduce delays, and streamline decision-making processes. This improves project efficiency, ensuring that sustainability goals are met within the allocated time and budget.

Stakeholder Satisfaction: Stakeholder engagement and effective communication build trust, enhance relationships, and improve stakeholder satisfaction. By involving stakeholders in sustainability discussions and incorporating their input, project teams can address their needs and expectations, resulting in a higher degree of stakeholder satisfaction.

Reputation and Market Advantage: Successful collaboration and communication in sustainable construction projects enhance the reputation of project teams and organizations. This can lead to increased market advantage, attracting clients and partners who value sustainability and trust in the team's ability to deliver sustainable outcomes.

Knowledge Sharing and Learning: Collaboration facilitates knowledge sharing and learning among project team members. By capturing and disseminating lessons learned, best practices, and successful sustainability strategies, project teams can continually improve their performance and contribute to the advancement of sustainable construction practices.

Project team collaboration and communication are fundamental for the successful implementation of sustainable construction practices. By aligning sustainability goals, integrating diverse expertise, and fostering effective communication, project teams can optimize sustainability outcomes. Strategies such as early engagement, clear roles and responsibilities, open decision-making processes, and stakeholder engagement are essential for effective collaboration. The benefits of project team collaboration and communication include enhanced sustainability performance, innovation, improved project efficiency, stakeholder satisfaction, reputation building, and knowledge sharing. By prioritizing effective collaboration and communication, construction professionals can drive positive change and contribute to a more sustainable built environment.

CHAPTER 9:
CASE STUDIES:
SUSTAINABLE
CONSTRUCTION
PROJECTS

9.1 High-Performance Green Buildings: Achieving Sustainability Excellence

High-performance green buildings represent the pinnacle of sustainable construction practices, integrating innovative design strategies, advanced technologies, and efficient building systems. These projects prioritize environmental stewardship, energy efficiency, occupant health, and overall sustainability performance. This section explores several case studies of high-performance green buildings, highlighting their key features, sustainable design strategies, and the benefits they bring.

Case Study 1: The Edge, Amsterdam, Netherlands

The Edge, located in Amsterdam, is widely regarded as one of the most sustainable office buildings in the world. It showcases cutting-edge sustainable design and technology, achieving the highest level of certification under the Leadership in Energy and Environmental Design (LEED) rating system. The building's sustainability features include:

Energy Efficiency: The Edge utilizes advanced energy management systems, including smart lighting, occupancy sensors, and a connected network of sensors to optimize energy consumption. The building's intelligent lighting system adjusts lighting levels based on natural daylight and occupancy, significantly reducing energy usage.

Sustainable Materials: The construction of The Edge prioritized the use of environmentally friendly materials with low embodied carbon. Sustainable timber, recycled materials, and non-toxic finishes were incorporated to minimize the building's environmental impact.

Indoor Environment Quality: The building prioritizes occupant health and comfort through features such as ample natural daylight, excellent air quality control, and thermal comfort optimization. The Edge's innovative climate ceiling system provides individual control of heating and cooling, ensuring a comfortable and productive working environment for occupants.

Renewable Energy Integration: The Edge incorporates renewable energy technologies, including solar panels and onsite energy generation systems. These systems generate a significant portion of the building's energy needs, contributing to its net-zero energy performance.

Case Study 2: One Central Park, Sydney, Australia

One Central Park in Sydney is an iconic example of sustainable urban development. The project combines residential and commercial spaces with extensive greenery, creating a harmonious blend of nature and architecture. Key sustainability features of One Central Park include:

Vertical Gardens: The building's façade incorporates extensive vertical gardens that not only enhance aesthetics but also provide natural shading, insulation, and improved air quality. The gardens feature a variety of plant species selected for their ability to thrive in an urban environment, contributing to the project's biodiversity.

Water Efficiency: One Central Park incorporates innovative water management strategies, including a blackwater treatment plant and rainwater harvesting systems. These systems supply water for non-potable uses such as toilet flushing and irrigation, significantly reducing the building's reliance on the municipal water supply.

Energy Innovation: The building utilizes a range of energy-efficient technologies, including a trigeneration system that generates electricity, heating, and cooling on-site. Solar panels and a heliostat system redirect sunlight to provide natural lighting to interior spaces, reducing the need for artificial lighting during daylight hours.

Public Space Integration: One Central Park features vibrant public spaces, including a central park and podium gardens, encouraging social interaction, community engagement, and a connection to nature. These green spaces not only provide aesthetic value but also contribute to improved air quality and overall well-being.

9.2 Net-Zero Energy Buildings: Achieving Energy Independence

Net-zero energy buildings (NZEBs) are designed to produce as much energy as they consume, resulting in a neutral or zero net energy usage over a defined period. These projects prioritize energy efficiency, renewable energy generation, and advanced building systems to achieve energy independence. Here are two notable case studies of net-zero energy buildings:

Case Study 1: Bullitt Center, Seattle, USA

The Bullitt Center in Seattle is a pioneering example of a net-zero energy commercial building. It showcases innovative sustainability features and serves as a model for sustainable urban development. Key sustainability features of the Bullitt Center include:

Energy Efficiency: The building incorporates passive design strategies, such as optimized daylighting, natural ventilation, and high-performance insulation, to minimize energy demand. Energy-efficient systems, including heat recovery ventilation and efficient lighting fixtures, further reduce energy consumption.

Renewable Energy Generation: The Bullitt Center generates its energy through a rooftop solar array, which provides electricity for the building's operations. The surplus energy generated is fed back into the grid, contributing to the local renewable energy supply.

Water Conservation: The building utilizes rainwater harvesting and greywater recycling systems to minimize water consumption. These systems collect rainwater for non-potable uses, such as toilet flushing and irrigation, reducing the demand for freshwater resources.

Healthy Indoor Environment: The Bullitt Center prioritizes occupant health and well-being through features such as ample daylight, low-toxicity materials, and excellent indoor air quality.

The building's operable windows allow for natural ventilation, enhancing occupant comfort and connection to the outdoors.

Case Study 2: Richardsville Elementary School, Kentucky, USA

Richardsville Elementary School is the first net-zero energy school in the United States, showcasing sustainable design and energy-efficient technologies in an educational setting. Key sustainability features of Richardsville Elementary School include:

Energy Monitoring and Optimization: The school incorporates advanced energy monitoring systems to track energy consumption in real-time. This data informs ongoing energy optimization efforts, allowing the school to fine-tune energy usage and identify areas for improvement.

Energy-Efficient Building Envelope: Richardsville Elementary School features a high-performance building envelope, including well-insulated walls, energy-efficient windows, and a reflective roof. These design elements minimize heat transfer and improve the building's thermal performance, reducing energy demands for heating and cooling.

Renewable Energy Integration: The school utilizes a combination of rooftop solar panels and geothermal heating and cooling systems to generate renewable energy on-site. The solar panels generate electricity, while the geothermal system utilizes the stable temperature of the earth to provide efficient heating and cooling throughout the year.

Educational Opportunities: Richardsville Elementary School incorporates sustainability education into its curriculum, providing students with hands-on learning experiences related to energy conservation, renewable energy, and sustainable practices. The building serves as a living laboratory, engaging students in understanding and embracing sustainable principles.

9.3 Adaptive Reuse and Renovation Projects: Transforming the Built Environment

Adaptive reuse and renovation projects involve the transformation of existing structures into sustainable and functional spaces. These projects embrace the principles of recycling, preservation, and resource efficiency, breathing new life into aging buildings. Here are two notable case studies of adaptive reuse and renovation projects:

Case Study 1: The Renewal of King's Cross Station, London, UK

The renewal of King's Cross Station in London exemplifies the successful transformation of an historic railway station into a modern, sustainable transportation hub. Key sustainability features of this project include:

Heritage Preservation: The project carefully preserved and restored the station's heritage features, respecting its architectural and historical significance. This approach ensures the retention of the building's cultural value while incorporating sustainable design elements.

Energy Efficiency: The renewal project implemented energy-efficient systems, such as LED lighting, efficient HVAC systems, and optimized insulation, to reduce energy consumption. The use of natural ventilation and daylighting strategies further enhances energy performance.

Sustainable Materials: The project incorporated sustainable materials and construction practices, including the use of recycled materials, low-impact finishes, and responsibly sourced timber. This approach minimized the environmental impact of the renovation and ensured the longevity of the building.

Accessibility and Public Transport Integration: The renewal of King's Cross Station prioritized accessibility, creating barrier-free access and improving connectivity to public transportation. Enhanced pedestrian pathways, cycling infrastructure, and efficient public transport connections encourage sustainable modes of travel.

Case Study 2: The High Line, New York City, USA

The High Line is a remarkable example of adaptive reuse, transforming an elevated railway into a vibrant public park and greenway. Key sustainability features of The High Line project include:

Brownfield Revitalization: The project involved the rejuvenation of a disused railway viaduct, transforming it into an elevated park that preserves the historical character of the structure while revitalizing the surrounding neighborhood. This approach reduces urban sprawl and promotes sustainable urban development.

Biodiversity Enhancement: The High Line incorporates extensive planting of native vegetation, creating a biodiverse habitat that attracts birds, butterflies, and other wildlife. The carefully curated plantings provide ecological benefits, support pollinators, and contribute to the overall resilience of urban ecosystems.

Stormwater Management: The park's design integrates stormwater management strategies, such as permeable paving and rain gardens, to capture and treat rainwater. These features help to mitigate stormwater runoff, reduce strain on the city's sewer system, and promote water conservation.

Community Engagement: The High Line project actively involved the local community in its planning and design process, ensuring that the park meets the needs and desires of its users. This approach fosters a sense of ownership and pride among community members, strengthening social cohesion and connection to the urban environment.

Cultural and Recreational Amenities: The adaptive reuse of the railway viaduct created a unique public space that offers cultural programming, art installations, and recreational opportunities. This approach enhances the quality of life for residents and visitors, promoting physical activity, leisure, and cultural engagement.

Case studies of sustainable construction projects, such as high-performance green buildings, net-zero energy buildings, and adaptive reuse projects, demonstrate the successful integration of sustainable design principles, innovative technologies, and resource-efficient strategies. These projects showcase the potential for achieving environmental, social, and economic sustainability in the built environment. By learning from these case studies, construction professionals can gain valuable insights and inspiration for their own sustainable construction endeavors, contributing to a more sustainable and resilient future.

CHAPTER 10: FINANCING AND ECONOMIC CONSIDERATIONS

10.1 Cost-Benefit Analysis of Sustainable Construction

C ost-benefit analysis plays a crucial role in evaluating the economic viability of sustainable construction projects. This analysis involves a systematic assessment of the costs incurred and the benefits gained from implementing sustainable practices. Understanding the financial implications and potential returns of sustainable construction is essential for decision-making, attracting investment, and driving the adoption of sustainable strategies. In this chapter, we explore the concept of cost-benefit analysis in the context of sustainable construction and highlight its importance in guiding project feasibility and financial decision-making.

The Importance of Cost-Benefit Analysis

Cost-benefit analysis provides a comprehensive understanding of the financial implications and potential returns associated with sustainable construction projects. Here's why it is important:

Project Feasibility: Cost-benefit analysis helps determine the financial feasibility of sustainable construction projects. By evaluating the costs and benefits over the project's lifecycle, decision-makers can assess whether the financial gains outweigh the investment costs, guiding project selection and prioritization.

Investment Attraction: Investors and financial institutions require a clear understanding of the financial viability of sustainable construction projects before committing resources. Cost-benefit analysis provides the necessary information to attract investment and secure financing for sustainability initiatives.

Risk Assessment: Cost-benefit analysis helps identify and assess potential risks and uncertainties associated with sustainable construction projects. It allows for a thorough evaluation of financial risks, including construction costs, operational expenses, and market risks, enabling project teams to develop risk mitigation strategies.

Performance Comparison: Cost-benefit analysis facilitates the comparison of different sustainability strategies and technologies. It helps decision-makers evaluate the financial impacts of various options, enabling informed choices about the most cost-effective and beneficial approaches to achieve sustainability goals.

Elements of Cost-Benefit Analysis

To conduct a robust cost-benefit analysis for sustainable construction projects, consider the following elements:

Project Costs: Identify and quantify all costs associated with sustainable construction, including design and planning expenses, construction materials, labor, equipment, permits, and certifications. It is essential to consider both upfront costs and long-term operational costs to capture the full financial implications.

Benefits: Determine and evaluate the tangible and intangible benefits of sustainable construction. Tangible benefits may include energy cost savings, reduced water consumption, and improved indoor air quality. Intangible benefits could involve enhanced occupant satisfaction, increased market value, and reputational advantages.

Timeframe and Discount Rates: Define the timeframe for the cost-benefit analysis, considering the project's lifecycle and relevant time horizons. Additionally, apply appropriate discount rates to account for the time value of money and reflect the project's risk profile.

Quantitative and Qualitative Assessment: Incorporate both quantitative and qualitative assessments of costs and benefits. While some aspects, such as energy savings, can be quantified accurately, others, like improved occupant well-being, may require qualitative assessments or survey-based data collection.

Sensitivity Analysis: Conduct sensitivity analyses to assess the impacts of uncertainties and variations in key assumptions on the project's financial outcomes. This analysis helps identify critical factors and potential risks that could significantly influence the cost-benefit assessment.

Financial Evaluation Tools

Various financial evaluation tools can assist in conducting cost-benefit analysis for sustainable construction projects. These tools include:

Net Present Value (NPV): NPV calculates the present value of expected cash inflows and outflows over the project's lifecycle, considering the discount rate. A positive NPV indicates that the project is financially viable, while a negative NPV suggests potential financial losses.

Return on Investment (ROI): ROI measures the profitability of an investment by comparing the project's net benefits to its costs. It is expressed as a percentage, with higher ROI indicating better financial performance.

Internal Rate of Return (IRR): IRR is the discount rate at which the project's net present value equals zero. It represents the project's average annual rate of return and helps assess its financial attractiveness.

Payback Period: Payback period determines the time it takes for the project's net benefits to recover the initial investment. A shorter payback period indicates quicker returns on investment.

Limitations and Considerations

When conducting cost-benefit analysis for sustainable construction projects, it is important to consider the following limitations and considerations:

Data Availability and Accuracy: Cost-benefit analysis heavily relies on accurate and reliable data. Availability of relevant data can sometimes pose challenges, particularly when assessing long-term benefits and intangible factors. Efforts should be made to collect robust data and use realistic assumptions to enhance the accuracy of the analysis.

Uncertainty and Risk: Cost-benefit analysis involves inherent uncertainties and risks. Changes in market conditions, technological advancements, and regulatory frameworks can impact the financial outcomes. Sensitivity analysis and risk assessment techniques should be employed to account for these uncertainties and evaluate their potential impacts.

Long-Term Perspective: Sustainable construction projects often involve long-term benefits that may extend beyond traditional project lifecycles. It is important to adopt a long-term perspective when assessing costs and benefits to capture the full value of sustainability investments.

Non-Monetary Considerations: Cost-benefit analysis should not solely focus on financial aspects. Non-monetary factors, such as social and environmental impacts, should also be considered. Incorporating qualitative assessments and multi-criteria evaluation techniques allows for a more comprehensive evaluation of sustainability outcomes.

Case Studies

To illustrate the application of cost-benefit analysis in sustainable construction, we will examine two case studies:

Case Study 1: Energy-Efficient Retrofit of a Commercial Building

In this case study, a commercial building undergoes an energy-efficient retrofit to reduce energy consumption and operational costs. The cost-benefit analysis considers upfront investment costs, energy savings over a 10-year period, and potential government incentives. The analysis reveals that the project has a positive NPV, indicating a financially viable investment. Additionally, the payback period is determined to be five years, demonstrating a reasonable return on investment.

Case Study 2: Green Building Certification for a Mixed-Use Development

In this case study, a mixed-use development seeks green building certification to enhance market value and attract tenants. The cost-benefit analysis compares the additional costs associated with green building features and certifications to the projected benefits, such as energy savings, reduced water consumption, and higher rental rates. The analysis shows that the project's positive NPV and attractive ROI justify the investment in sustainable features, considering both financial gains and intangible benefits.

Cost-benefit analysis is a valuable tool for evaluating the economic viability and financial implications of sustainable construction projects. It helps decision-makers assess project feasibility, attract investment, and prioritize sustainability initiatives. By considering project costs, tangible and intangible benefits, and employing appropriate financial evaluation tools, construction professionals can make informed decisions that balance economic considerations with sustainability goals. When conducted rigorously, cost-benefit analysis facilitates the transition towards a more sustainable built environment, delivering positive financial outcomes and long-term value.

10.2 Financial Incentives and Green Financing Options
The adoption of sustainable construction practices is often supported by a range of financial incentives and green financing options. These mechanisms aim to incentivize and facilitate sustainable development by reducing financial barriers, promoting investments in environmentally friendly projects, and offering favorable terms and conditions. In this chapter, we explore the various financial incentives and green financing options available for sustainable construction projects and discuss their significance in driving the transition towards a more sustainable built environment.

Financial Incentives for Sustainable Construction

Financial incentives are provided by governments, municipalities, and organizations to promote sustainable construction practices. These incentives can take different forms, including grants, subsidies, tax credits, and rebates. Here are some common financial incentives available:

Tax Incentives: Governments often offer tax incentives to encourage sustainable construction. These may include tax credits for energy-efficient buildings, renewable energy systems, or green building certifications. Tax incentives reduce the tax burden on developers and investors, making sustainable projects more financially attractive.

Grants and Subsidies: Governments and organizations provide grants and subsidies to support sustainable construction initiatives. These funds can be used to offset upfront costs, such as energy audits, feasibility studies, and the implementation of sustainable technologies. Grants and subsidies help reduce financial barriers and promote the adoption of sustainable practices.

Green Building Certification Incentives: Many green building certification programs offer incentives to encourage developers to pursue certification. These incentives can include reduced certification fees, expedited permit processes, and marketing support. By rewarding sustainable achievements, these programs motivate project teams to prioritize sustainable construction practices.

Energy Performance Contracts: Energy performance contracts involve agreements between building owners and energy service companies (ESCOs) to improve energy efficiency and reduce operating costs. ESCOs finance, implement, and maintain energy-saving measures, and the cost savings from improved efficiency are shared between the parties. This financing model allows building owners to achieve sustainability goals without upfront capital investment.

Feed-in Tariffs and Power Purchase Agreements (PPAs): Feed-in tariffs and PPAs are mechanisms that provide financial incentives for renewable energy generation. These arrangements guarantee a fixed price for renewable energy generated by a building or project, incentivizing the implementation of solar panels, wind turbines, or other renewable energy systems.

Green Financing Options for Sustainable Construction

Green financing options enable the financing of sustainable construction projects by offering specialized loan products, investment funds, and financial instruments. These options recognize the long-term value of sustainable investments and provide favorable terms and conditions. Here are some common green financing options:

Green Loans: Green loans are specifically designed for sustainable projects and offer favorable terms to support their implementation. These loans may have lower interest rates, longer repayment periods, and flexible repayment structures. Green loans are typically available for projects that meet certain sustainability criteria, such as green building certifications or energy-efficient designs.

Green Bonds: Green bonds are debt instruments issued by governments, municipalities, or corporations to finance environmentally friendly projects. The proceeds from these bonds are earmarked for sustainable initiatives, including sustainable construction projects. Investors are attracted to green bonds due to their alignment with environmental objectives and potential financial returns.

Energy Efficiency Financing: Energy efficiency financing programs provide loans or financing options specifically for energy-efficient upgrades and retrofits. These programs may offer favorable terms, such as lower interest rates or longer repayment periods, to encourage building owners to invest in energy-saving measures.

Property-Assessed Clean Energy (PACE) Financing: PACE financing allows property owners to finance energy efficiency improvements through a special assessment on their property tax bill. The financing is repaid over an extended period, and the loan is attached to the property rather than the owner. PACE financing makes it easier for property owners to access capital for sustainability upgrades without relying on traditional loans.

Impact Investment Funds: Impact investment funds focus on projects that generate positive social and environmental impacts alongside financial returns. These funds invest in sustainable construction projects that align with their sustainability objectives, providing capital and expertise to support their implementation.

Evaluating Return on Investment for Sustainable Projects

Evaluating the return on investment (ROI) for sustainable construction projects is crucial to assess their financial viability and attractiveness. ROI analysis compares the project's financial gains to the costs incurred, providing insights into the profitability and payback period of the investment. Here are key considerations for evaluating ROI:

Financial Metrics: ROI analysis involves financial metrics such as net present value (NPV), internal rate of return (IRR), payback period, and profitability index. These metrics quantify the financial performance of the project and help assess its attractiveness.

Cost Savings: Sustainable construction projects often result in long-term cost savings due to reduced energy consumption, water usage, and operational expenses. Evaluating the potential cost savings over the project's lifecycle is essential in determining the project's ROI.

Additional Benefits: Beyond financial returns, sustainable construction projects offer additional benefits that contribute to ROI. These may include enhanced market value, improved occupant health and productivity, reduced environmental impact, and increased stakeholder satisfaction.

Risk Analysis: ROI analysis should consider potential risks and uncertainties associated with sustainable construction projects. These risks may include changes in regulations, market demand, technology advancements, and project-specific factors. Conducting sensitivity analysis and risk assessment helps evaluate the project's resilience and potential impacts on ROI.

Financial incentives and green financing options are instrumental in unlocking sustainable investments and promoting the adoption of sustainable construction practices. These mechanisms provide financial support, reduce barriers, and incentivize developers and investors to prioritize sustainability. Through cost-saving incentives, specialized financing products, and favorable terms, financial institutions and governments facilitate the transition towards a more sustainable built environment. Evaluating ROI for sustainable projects ensures sound financial decision-making, considering both financial gains and the broader benefits that contribute to a sustainable future. By leveraging financial incentives and green financing options and conducting robust ROI analysis, construction professionals can drive positive change and contribute to a more sustainable and resilient built environment.

10.3 Evaluating Return on Investment for Sustainable Projects

Evaluating the return on investment (ROI) for sustainable projects is a critical aspect of financial decision-making. ROI analysis assesses the financial performance and profitability of a sustainable construction project, taking into account the costs incurred and the benefits gained over its lifecycle. By evaluating the ROI, construction professionals can determine the economic viability and attractiveness of sustainable investments. In this chapter, we delve into the process of evaluating ROI for sustainable projects, highlighting key considerations and methodologies.

Understanding Return on Investment (ROI)

Return on investment is a financial metric that measures the profitability of an investment relative to its cost. It quantifies the percentage return gained from the investment over a specified period. For sustainable construction projects, ROI analysis provides insights into the financial returns generated from implementing sustainable practices, technologies, and design features.

Key Considerations for Evaluating ROI

When evaluating ROI for sustainable projects, several key considerations should be taken into account:

Project Costs: Identify and quantify all costs associated with the sustainable construction project, including upfront investment costs, design and planning expenses, construction and material costs, certification fees, and ongoing operational expenses. It is crucial to capture both the direct and indirect costs comprehensively.

Financial Benefits: Evaluate the financial benefits generated by the sustainable project. These benefits may include energy savings, reduced water consumption, lower operational and maintenance costs, increased property value, and potential revenue streams from renewable energy generation or carbon credits. Consider both tangible and intangible financial benefits. **Timeframe**: Define the timeframe for ROI evaluation, considering the project's lifecycle and relevant time horizons. Sustainable projects often involve long-term benefits that may extend beyond traditional investment evaluation periods. It is important to adopt a timeframe that captures the full financial impact of the project.

Discount Rate: Apply an appropriate discount rate to account for the time value of money. The discount rate reflects the opportunity cost of investing capital elsewhere. It is used to calculate the present value of future cash flows and determine the net present value (NPV) of the project.

Risk Assessment: Consider the risks associated with the sustainable project and their potential impact on the ROI. These risks may include market uncertainties, regulatory changes, technology advancements, and project-specific factors. Conduct a risk assessment to identify and mitigate potential risks and uncertainties.

Methodologies for ROI Analysis

Several methodologies can be employed to evaluate ROI for sustainable projects:

Net Present Value (NPV): NPV analysis calculates the present value of the project's expected cash inflows and outflows, discounted at the appropriate rate. A positive NPV indicates that the project is financially viable and generates more value than the initial investment. A negative NPV suggests potential financial losses.

Internal Rate of Return (IRR): IRR is the discount rate at which the NPV of the project becomes zero. It represents the average annual rate of return generated by the investment. A higher IRR indicates better financial performance and attractiveness.

Payback Period: The payback period measures the time required for the project's cash inflows to recoup the initial investment. It is the duration within which the project becomes financially self-sufficient. A shorter payback period indicates a quicker return on investment.

Sensitivity Analysis: Conduct sensitivity analysis to assess the impact of varying key assumptions, such as construction costs, energy prices, or occupancy rates, on the project's financial outcomes. This analysis helps identify the critical factors that significantly influence the ROI and provides insights into the project's resilience to changes.

Non-Financial Considerations

While ROI analysis primarily focuses on financial performance, it is important to consider non-financial factors that contribute to the overall value of sustainable projects. These non-financial considerations may include environmental benefits, social impact, improved occupant health and productivity, brand reputation, and regulatory compliance. Assessing these factors alongside financial metrics provides a more comprehensive evaluation of the project's value.

Case Studies

To illustrate the evaluation of ROI for sustainable projects, let's examine two case studies:

Case Study 1: Energy-Efficient Retrofit of an Office Building

In this case study, an office building undergoes an energy-efficient retrofit that includes LED lighting, HVAC system upgrades, and enhanced insulation. The ROI analysis considers the upfront costs of the retrofit, energy savings over a 10-year period, and potential government incentives. The analysis reveals a positive NPV and an attractive payback period of six years, indicating a financially viable investment with long-term cost savings.

Case Study 2: Integration of Solar Power in a Residential Development

In this case study, a residential development incorporates solar power systems into its design. The ROI analysis assesses the installation costs, energy generation potential, and long-term energy cost savings. Additionally, it considers the potential revenue from feed-in tariffs or net metering programs. The analysis reveals a positive NPV, a favorable IRR, and a reasonable payback period, demonstrating the financial viability of the solar power integration.

Evaluating ROI for sustainable construction projects is essential for making informed financial decisions and determining the economic viability of investments. By considering project costs, financial benefits, timeframe, discount rates, and conducting sensitivity analysis, construction professionals can assess the financial performance and attractiveness of sustainable projects. It is important to complement ROI analysis with the consideration of non-financial factors to capture the full value of sustainability initiatives. Through robust ROI analysis, construction professionals can confidently prioritize sustainable investments, contributing to a more sustainable and resilient built environment while generating favorable financial returns.

CHAPTER 11:
BUILDING RESILIENCE
AND CLIMATE
ADAPTATION

11.1 Designing for Extreme Weather Events

Designing for extreme weather events is crucial in creating resilient built environments that can withstand and adapt to the challenges posed by climate change. This chapter explores various strategies and considerations for designing buildings and infrastructure to mitigate the impacts of extreme weather events such as hurricanes, floods, heatwaves, and wildfires. By incorporating resilient design principles, construction professionals can enhance the safety, functionality, and longevity of structures in the face of changing climatic conditions.

Understanding Extreme Weather Events

Extreme weather events, intensified by climate change, have become more frequent and severe in recent years. These events include hurricanes, storms, heavy rainfall, heatwaves, droughts, and wildfires. Understanding the characteristics and potential impacts of these events is vital for designing resilient structures. Factors such as wind loads, flooding potential, temperature extremes, and fire risks must be considered during the design process.

Designing for Resilience

Designing for resilience involves integrating strategies that enhance the durability, adaptability, and safety of buildings and infrastructure. Here are key considerations for designing resilient structures:

Robust Structural Systems: Employing robust structural systems that can withstand the forces exerted by extreme weather events is essential. This includes using resilient materials, designing for appropriate load capacities, and considering potential hazards such as wind, water, or seismic forces.

Site Selection and Planning: Careful site selection and planning can help mitigate the impacts of extreme weather events. Avoiding flood-prone areas, considering the elevation and topography of the site, and incorporating natural barriers can minimize risks and enhance resilience.

Stormwater Management: Implementing effective stormwater management systems is crucial in areas prone to heavy rainfall and flooding. This involves the design of proper drainage systems, retention ponds, and the use of permeable surfaces to reduce runoff and prevent water damage.

Climate-Responsive Building Envelope: A climate-responsive building envelope minimizes the transfer of heat, cold, and moisture, enhancing thermal comfort and energy efficiency. It should be designed to withstand temperature extremes, moisture infiltration, and wind-driven rain.

Wind and Seismic Resistance: Designing structures to resist high wind loads and seismic forces is vital in areas prone to hurricanes, tornadoes, and earthquakes. This includes proper structural detailing, anchorage systems, and reinforcement to enhance the building's resistance to lateral forces.

Heat Mitigation: Incorporating strategies to mitigate heat stress is crucial, particularly in regions experiencing heatwaves. This includes the use of shading devices, reflective materials, natural ventilation systems, and cool roof technologies to reduce heat absorption and improve thermal comfort.

Climate-Responsive Building Systems

Climate-responsive building systems are designed to adapt to changing climatic conditions and optimize energy efficiency. Here are some key considerations for climate-responsive design:

Energy-Efficient HVAC Systems: Implementing energy-efficient heating, ventilation, and air conditioning (HVAC) systems can reduce energy consumption and enhance indoor comfort. This includes the use of high-efficiency equipment, zoned controls, and smart building automation systems.

Natural Ventilation: Incorporating natural ventilation strategies allows for the passive exchange of air, reducing the reliance on mechanical systems and improving indoor air quality. This includes the strategic placement of windows, operable louvers, and the use of stack effect for natural airflow.

Daylighting and Lighting Controls: Maximizing natural daylight and incorporating lighting controls can reduce energy usage for artificial lighting while enhancing occupant well-being. This involves optimizing window sizes and locations, using daylight sensors, and employing efficient lighting fixtures.

Renewable Energy Integration: Integrating renewable energy systems, such as solar panels or wind turbines, can enhance the resilience of buildings and reduce reliance on fossil fuels. This includes evaluating the feasibility of renewable energy generation based on site conditions and energy demands.

Enhancing Resilience in Construction Practices

Enhancing resilience in construction practices involves adopting approaches that ensure the durability, adaptability, and safety of buildings. Here are some key considerations:

Robust Construction Techniques: Employing robust construction techniques, such as proper waterproofing, reinforcement detailing, and quality control measures, ensures the longevity and structural integrity of buildings.

Sustainable Materials Selection: Choosing durable and sustainable materials that can withstand extreme weather events is crucial. This includes considering factors such as resistance to moisture, fire, and wind, as well as the use of recycled or locally sourced materials to reduce environmental impact.

Post-Disaster Recovery Planning: Developing post-disaster recovery plans and strategies is essential to facilitate the swift and efficient recovery of communities and buildings after extreme weather events. This involves establishing emergency response procedures, backup systems, and resilient infrastructure.

Community Engagement and Education: Engaging and educating communities about the importance of resilient construction practices is vital. This includes raising awareness about potential risks, providing guidance on emergency preparedness, and promoting sustainable building practices.

Case Studies

To illustrate the implementation of resilient design principles, let's examine three case studies:

Case Study 1: Hurricane-Resistant Building Design

In this case study, a coastal community implements hurricane-resistant design strategies for residential and commercial buildings. These strategies include elevated foundations, impact-resistant windows and doors, reinforced structural systems, and resilient roofing materials. The design ensures that the buildings can withstand high wind speeds, storm surges, and flying debris associated with hurricanes.

Case Study 2: Climate-Responsive Office Building

In this case study, an office building incorporates climate-responsive design features to adapt to temperature extremes and reduce energy consumption. The building envelope is designed to minimize heat gain and loss, and energy-efficient HVAC systems are installed to provide thermal comfort. Natural ventilation strategies and daylighting techniques are also employed to enhance occupant well-being and reduce reliance on mechanical systems.

Case Study 3: Flood-Resilient Infrastructure

In this case study, a city implements flood-resilient infrastructure to mitigate the impacts of heavy rainfall and flooding. The design incorporates elevated roads, flood walls, and stormwater management systems to redirect and manage water flow. Flood-resistant materials and finishes are used in buildings, and critical infrastructure is protected against water damage.

Designing for extreme weather events, implementing climate-responsive building systems, and enhancing resilience in construction practices are essential in creating built environments that can withstand and adapt to the challenges of climate change. By incorporating resilient design principles, construction professionals can enhance the safety, functionality, and longevity of structures. Through careful site selection, robust structural systems, climate-responsive building systems, and community engagement, construction professionals play a pivotal role in creating resilient communities. By embracing resilient design, we can build a more sustainable and resilient future in the face of climate uncertainties and extreme weather events.

11.2 Climate-Responsive Building Systems

In the face of climate change, constructing buildings with climate-responsive systems is crucial for enhancing energy efficiency, occupant comfort, and overall resilience. Climate-responsive building systems adapt to the local climate and weather conditions, optimizing performance while reducing environmental impact. This chapter explores various climate-responsive building systems and strategies that construction professionals can employ to create sustainable and adaptive structures.

Passive Design Strategies

Passive design strategies maximize the use of natural elements to regulate temperature, humidity, and ventilation within a building. By incorporating these strategies, construction professionals can reduce the reliance on mechanical systems and minimize energy consumption. Key passive design strategies include:

Orientation and Building Form: Proper building orientation and form take advantage of natural sunlight, shading, and prevailing winds. Orienting the building to maximize solar gain in winter and minimize it in summer can enhance energy efficiency and occupant comfort.

Building Envelope: The building envelope plays a vital role in regulating heat transfer and insulation. Effective insulation, high-performance windows, and air sealing techniques minimize heat gain or loss, reduce energy demands for heating and cooling, and enhance thermal comfort.

Natural Ventilation: Natural ventilation leverages the movement of air to provide cooling and fresh air exchange. Incorporating operable windows, louvers, and stack effect principles allows for the passive flow of air, reducing the need for mechanical ventilation systems.

Daylighting: Maximizing natural daylight through strategic window placement and design improves visual comfort, reduces reliance on artificial lighting, and positively impacts occupant well-being and productivity.

Active Building Systems

Active building systems utilize mechanical and electrical equipment to regulate indoor environmental conditions efficiently. These systems work in harmony with passive design strategies to optimize comfort and energy performance. Key climate-responsive active building systems include:

Heating, Ventilation, and Air Conditioning (HVAC): Energy-efficient HVAC systems maintain comfortable indoor temperatures while minimizing energy consumption. Advanced technologies, such as variable refrigerant flow systems, heat recovery ventilation, and demand-controlled ventilation, enable precise temperature and air quality control.

Lighting Systems: Energy-efficient lighting systems, such as LED technology and smart lighting controls, minimize energy consumption and provide flexibility in adjusting lighting levels based on occupancy and daylight availability.

Renewable Energy Integration: Integrating renewable energy sources, such as solar photovoltaic systems or wind turbines, reduces reliance on fossil fuel-based electricity and lowers carbon emissions. These renewable energy systems can power the building's active systems and contribute to overall energy sustainability.

Building Automation and Controls: Advanced building automation and control systems monitor and regulate various building systems, optimizing their performance based on real-time data. These systems enable precise control, scheduling, and energy management, ensuring efficient operation of active building systems.

Energy Management and Efficiency

Effective energy management and efficiency measures are crucial for climate-responsive buildings. Construction professionals can implement the following strategies:

Energy Monitoring and Benchmarking: Monitoring and

benchmarking energy consumption provide valuable insights into building performance, enabling identification of areas for improvement and optimization.

Energy-Efficient Appliances and Equipment: Selecting energy-efficient appliances and equipment, such as ENERGY STAR-rated devices, helps reduce overall energy consumption and operational costs.

Demand Response Strategies: Implementing demand response strategies allows buildings to respond to peak electricity demand and grid conditions by temporarily reducing energy consumption or shifting loads to off-peak hours. This supports grid stability and reduces stress on the energy infrastructure.

Energy Storage Solutions: Integrating energy storage systems, such as batteries, allows buildings to store excess renewable energy generated during low-demand periods for later use. This enhances energy self-sufficiency and supports load balancing.

Smart Building Technologies

Leveraging smart building technologies enhances the adaptive capabilities of climate-responsive systems. These technologies enable real-time monitoring, data analysis, and automated control, optimizing building performance and occupant comfort. Key smart building technologies include:

Building Energy Management Systems (BEMS): BEMS enable centralized monitoring, control, and optimization of various building systems, such as HVAC, lighting, and renewable energy systems. BEMS utilize data analytics and advanced algorithms to optimize energy usage and reduce operational costs.

Internet of Things (IoT) Sensors: IoT sensors collect data on occupancy, temperature, humidity, and other environmental factors. This data informs the control systems and enables automated adjustments to optimize energy consumption and occupant comfort.

Occupant Engagement and Feedback Systems: Engaging occupants through feedback systems and user-friendly interfaces empowers them to actively participate in energy-saving behaviors. Real-time feedback on energy consumption encourages energy-conscious actions and fosters a culture of sustainability.

Case Studies

To illustrate the implementation of climate-responsive building systems, let's examine two case studies:

Case Study 1: Energy-Efficient Office Building

In this case study, an office building incorporates climate-responsive building systems, including passive design strategies, energy-efficient HVAC systems, smart lighting controls, and renewable energy integration. The combination of these systems optimizes energy performance, improves occupant comfort, and reduces environmental impact.

Case Study 2: Green Residential Development

In this case study, a residential development employs climate-responsive building systems, such as natural ventilation, high-performance building envelope, and demand response strategies. The integration of these systems ensures energy efficiency, promotes occupant well-being, and aligns with sustainability goals.

Climate-responsive building systems play a vital role in creating sustainable, energy-efficient, and adaptive structures. By combining passive design strategies, active building systems, energy management measures, and smart building technologies, construction professionals can optimize the performance and resilience of buildings in the face of changing environmental conditions. Implementing these systems not only reduces energy consumption and operational costs but also enhances occupant comfort and contributes to a more sustainable built environment. By embracing climate-responsive building systems, construction professionals lead the way in creating resilient structures that mitigate climate change impacts and support a more sustainable future.

11.3 Enhancing Resilience in Construction Practices

Enhancing resilience in construction practices is vital for creating buildings and infrastructure that can withstand and recover from various challenges, including climate change, natural disasters, and societal disruptions. This chapter explores key strategies and considerations for enhancing resilience in construction, ensuring the durability, adaptability, and safety of built environments.

Risk Assessment and Planning

Identify and Assess Risks: Conduct a comprehensive risk assessment to identify potential hazards and vulnerabilities that the construction project may face. Consider natural disasters, climate-related risks, social and economic disruptions, and technological risks. This assessment will help determine the necessary resilience measures.

Resilience Planning: Develop a resilience plan that outlines the strategies and actions to mitigate identified risks. This plan should include measures for hazard mitigation, emergency response, business continuity, and post-disaster recovery. Engage relevant stakeholders and collaborate with local authorities to align with existing resilience initiatives.

Robust Structural Design and Construction

Resilient Structural Systems: Implement resilient structural design principles that can withstand extreme loads and environmental conditions. Consider wind, seismic, and flood-resistant designs, as well as proper foundation systems and reinforcement. Use durable materials and construction techniques that can resist wear, deterioration, and corrosion.

Building Envelope Protection: Ensure a resilient building envelope that can resist moisture intrusion, air leakage, and thermal bridging. Utilize high-quality materials, effective insulation, and proper sealing techniques to maintain energy efficiency, indoor comfort, and prevent damage from water infiltration.

Resilient Roofing and Exterior Elements: Install resilient roofing systems that can withstand strong winds, heavy rainfall, and hail. Incorporate protective features such as impact-resistant materials, secure flashings, and proper drainage systems. Consider resilient exterior elements, such as cladding and windows, to enhance durability and weather resistance.

Fire Resistance: Implement fire-resistant materials, construction techniques, and active fire protection systems to minimize the risk and impact of fires. Incorporate fire-resistant barriers, smoke control systems, and adequate fire detection and suppression systems to protect occupants and property.

Incorporating Adaptability and Flexibility

Modular and Flexible Design: Embrace modular and flexible design approaches that allow for future adaptation and modification. This enables buildings and infrastructure to respond to changing needs, technological advancements, and unforeseen circumstances. Design spaces that can easily accommodate changes in function, layout, or occupancy requirements.

Multi-Use and Redundancy: Plan for multi-use spaces that can serve different purposes or functions, promoting flexibility and resilience. Consider redundant systems, such as backup power supplies and redundant HVAC systems, to ensure continued operation during disruptions.

Future-Proofing: Anticipate future challenges and changes in climate, technology, and regulations when designing and constructing buildings. Incorporate measures that future-proof the built environment, such as integrating smart technologies, sustainable design features, and adaptable infrastructure.

Resilient Infrastructure Systems

Utility Resilience: Enhance the resilience of utility systems, including water, electricity, telecommunications, and wastewater management. This may involve measures such as underground utility placement, redundant systems, and backup power generation to ensure continuity of services during disruptions.

Sustainable Drainage Systems: Implement sustainable drainage systems, such as rain gardens, permeable pavements, and retention ponds, to manage stormwater effectively. These systems help mitigate flood risks, reduce strain on municipal infrastructure, and promote water conservation.

Green Infrastructure: Incorporate green infrastructure elements, such as green roofs, bioswales, and urban forests, to enhance stormwater management, mitigate heat island effects, and promote biodiversity. These features contribute to the overall resilience and sustainability of the built environment.

Collaboration and Stakeholder Engagement

Engage Local Communities: Involve local communities, stakeholders, and end-users in the construction process. Seek their input, address their concerns, and incorporate their needs and aspirations into the design and construction decisions. Engaging communities fosters a sense of ownership, improves resilience, and ensures that infrastructure serves the community's long-term interests.

Collaboration with Authorities and Experts: Collaborate with local authorities, emergency management agencies, and relevant experts to ensure compliance with regulations and incorporate best practices for resilience. Tap into their knowledge and expertise to inform design choices, construction techniques, and emergency response planning.

Knowledge Sharing and Training: Promote knowledge sharing and capacity building among construction professionals, contractors, and workers. Provide training on resilient construction practices, hazard awareness, and emergency response procedures. This ensures that all stakeholders are equipped with the necessary skills and knowledge to enhance resilience in construction.

Case Studies

To illustrate the implementation of resilient construction practices, let's examine three case studies:

Case Study 1: Resilient Hospital Design

In this case study, a hospital is designed and constructed with resilient features to withstand natural disasters and ensure uninterrupted healthcare services. The design incorporates flood-resistant materials, backup power systems, secure data storage, and redundant medical equipment. These features enhance the hospital's ability to provide critical services during emergencies.

Case Study 2: Resilient Infrastructure in Coastal Communities

In this case study, coastal communities implement resilient infrastructure measures to adapt to rising sea levels and increased storm surges. The construction of elevated roadways, flood barriers, and stormwater management systems mitigates flood risks, protects critical infrastructure, and ensures the continued functioning of the communities during extreme weather events.

Case Study 3: Resilient Commercial Building Retrofit

In this case study, a commercial building undergoes a retrofit to enhance its resilience to earthquakes. The retrofit includes strengthening the structural system, implementing flexible interior layouts, and installing seismic isolation devices. These measures improve occupant safety, protect property, and enable business continuity after seismic events.

Enhancing resilience in construction practices is essential for creating a sustainable, robust, and adaptable built environment. By conducting risk assessments, employing resilient design and construction techniques, incorporating adaptability and flexibility, and engaging stakeholders, construction professionals can contribute to the development of resilient communities. Through collaboration, knowledge sharing, and the implementation of best practices, the construction industry plays a crucial role in ensuring the durability, adaptability, and safety of buildings and infrastructure in the face of various challenges. By embracing resilience, we create a more sustainable and resilient future for generations to come.

Conclusion

The future of sustainable construction is marked by a growing emphasis on environmentally responsible practices, resource efficiency, and resilience. As awareness of climate change and environmental degradation increases, sustainable construction will become a standard approach, with a focus on reducing carbon emissions, conserving resources, and creating healthy and resilient built environments.

Emerging technologies and innovations are driving the transformation of the construction industry towards sustainability. Advancements in renewable energy systems, smart building technologies, artificial intelligence, building information modeling (BIM), and off-site construction methods are revolutionizing how buildings are designed, constructed, and operated. These innovations offer opportunities to enhance energy efficiency, optimize resource management, and improve the overall performance and sustainability of buildings.

Sustainable construction is a global endeavor that requires

understanding and incorporating diverse perspectives. Different regions and countries have unique environmental, social, and economic contexts that shape their approach to sustainable construction. Global perspectives on sustainable construction involve learning from and sharing best practices, considering local challenges and opportunities, and collaborating across borders to address global sustainability goals.

The transition to sustainable construction presents both opportunities and challenges. The opportunities include the growing demand for sustainable buildings and infrastructure, cost savings through energy efficiency, the development of green jobs and skills, and the potential for positive social and environmental impacts. However, challenges such as upfront costs, limited awareness and education, resistance to change, and the need for supportive policies and market incentives must be overcome. Embracing the opportunities while addressing the challenges is crucial for the continued progress of sustainable construction.

The future of sustainable construction is characterized by a focus on environmental responsibility, resource efficiency, and resilience. Emerging technologies and innovations play a vital role in driving this transformation, while global perspectives ensure a holistic and inclusive approach. The opportunities and challenges ahead require collaboration, innovation, and a commitment to sustainable practices. By embracing these aspects, the construction industry can pave the way for a more sustainable and resilient built environment that benefits present and future generations.

ABOUT THE AUTHOR

Steven Smith

Steven Smith is a renowned expert in the field of Construction Management, with a wealth of knowledge and experience spanning both academia and industry. With a doctorate in Construction Management from a prestigious university, Smith has dedicated his career to advancing the field and contributing to its body of knowledge.

Smith's academic journey began with a passion for understanding the intricacies of construction processes and finding innovative solutions to the challenges faced in the industry. His doctoral research focused on optimizing project management practices and enhancing productivity in construction projects. Through rigorous study and extensive research, Smith has developed a deep understanding of the various aspects of construction management and their impact on project success.

In addition to his academic achievements, Smith has also gained significant industry experience, working on various construction projects in collaboration with leading construction companies. This hands-on experience has provided him with valuable insights into the practical aspects of construction management, enabling him to bridge the gap between theory and real-world application.

Smith's contributions to the field of construction management extend beyond his academic pursuits. He is a prolific author, having published numerous articles in reputable journals and presented his research findings at international conferences. His work has explored diverse topics such as construction productivity, health and safety, and sustainable construction practices. Smith's publications have made a significant impact on the industry, helping professionals and scholars alike to enhance their understanding and improve their practices in construction management.

As a highly regarded academic and mentor, Smith has guided and inspired countless students pursuing careers in construction management. His dedication to education and his ability to convey complex concepts in a clear and accessible manner have earned him the respect and admiration of his students.

Throughout his career, Smith has demonstrated a commitment to continuous learning and professional development. In his spare time, Smith enjoys traveling to construction sites, exploring new construction technologies, and engaging in conversations with industry professionals to stay at the forefront of industry trends.

With his vast knowledge, extensive experience, and dedication to the field, Steven Smith brings a wealth of expertise to his role as an author, ensuring that his book provides valuable insights and practical guidance to construction professionals, academics, and students alike.

BOOKS BY THIS AUTHOR

Aerogel As A Sustainable Construction Material: Towards Net Zero Carbon Emissions

This book is a comprehensive guide for anyone interested in learning about a groundbreaking material (Aerogel) and its potential to revolutionize sustainable building practices. The book provides a detailed overview of aerogel, its unique properties, and its applications in the construction industry. It explains how aerogel can be used to promote energy efficiency and reduce carbon emissions, making it an essential tool for achieving net-zero carbon footprints in construction projects.

Written by an expert in the field, this book covers a range of topics, including the science behind aerogel, its production and manufacturing, and its use in different building applications. It also includes case studies and real-world examples that showcase the practical applications of aerogel in sustainable construction.

The book is recommended for construction professionals, researchers, students, and people that are simply interested in learning about the latest developments in sustainable building practices. With its clear and accessible writing style, engaging content, and practical insights, the book is sure to inspire and inform readers about the potential of aerogel as a game-changing material in the pursuit of sustainable construction practices.

The Dictionary Of Construction Terminologies: A Compendium Of Knowledge For Students, Academics, Practitioners, And House Owners

The dictionary of construction terminologies book is a comprehensive reference guide that provides definitions and explanations of the technical language and jargon used in the construction industry. It is an invaluable resource for professionals working in construction, as well as for students learning about the industry or for individuals looking to understand construction-related concepts better.

The book features a wide range of entries that cover various aspects of construction, including architecture, engineering, materials, equipment, and techniques. The book also provides clear and concise definitions of technical terms, written in easy-to-understand language. Terminologies are presented alphabetically to help readers find the descriptions they need quickly and easily. Whether you are a professional working in the field or interested in construction, this book is an essential tool to help you navigate the complex world of construction terminology with confidence and clarity.

Are you a student of construction, a house owner, an academic in the construction industry, or a practitioner that desires to acquire more knowledge about construction terms? If your answer to the preceding question is affirmative, this book may be one of the best investments you will ever make.

A Practical Approach To Risk Management In Construction: Unlocking Project Success

Discover the essential guide to practical risk management in the construction industry. This comprehensive book provides construction professionals with the knowledge and tools they need to navigate the complex world of risks and ensure project success. From identifying and analyzing risks to developing effective mitigation strategies, this book offers practical techniques and real-world examples that will empower you to proactively manage risks and deliver exceptional projects.

By delving into the critical aspects of risk management, this book equips you with the skills to anticipate and address potential challenges throughout the project lifecycle. You'll learn how to incorporate risk assessment into project planning, engage stakeholders in risk identification, and implement robust risk response strategies. The book also explores the integration of sustainability considerations into risk management, showcasing the industry's evolving focus on environmentally responsible practices.

Through engaging case studies and lessons learned from real-life scenarios, you'll gain invaluable insights into the application of risk management in diverse construction projects. The book emphasizes the importance of practical approaches and provides actionable strategies for addressing risks effectively, maximizing resource utilization, and fostering collaboration among project stakeholders.

Technology-driven risk analysis tools, such as data analytics and artificial intelligence, are also explored, demonstrating how these innovations can enhance risk management outcomes and decision-making. Furthermore, the book emphasizes the significance of continuous improvement, knowledge-sharing, and ethical leadership in creating a risk-aware culture within construction organizations.

Whether you're a project manager, contractor, or construction professional, this book will be your indispensable companion in navigating the complex landscape of construction risks.

By implementing the practical techniques and best practices outlined in this book, you'll be equipped to proactively manage risks, mitigate uncertainties, and deliver successful and resilient projects.

The Art Of Construction Project And Business Management

This resource is a comprehensive and indispensable guide for professionals in the construction industry. The groundbreaking book unravels the intricacies of managing construction projects and businesses with precision, expertise, and a strategic approach.

Drawing on years of industry experience and the latest industry insights, this book takes you on a transformative journey through every stage of the construction project lifecycle. From project initiation to execution, control, and beyond, you'll gain a deep understanding of the key principles, best practices, and real-world challenges faced by construction project managers and business leaders.

Unlike any other resource available, this book goes beyond theoretical concepts, offering practical, actionable strategies that can be immediately applied in your daily work. With its meticulous attention to detail, it equips you with the skills, knowledge, and tools to achieve project success, drive business growth, and navigate the ever-evolving landscape of the construction industry.

Inside, you'll explore essential topics such as project planning, scheduling, and risk management, ensuring that your projects are meticulously crafted for success. Learn how to navigate complex legal and regulatory frameworks, while maintaining the highest ethical standards and corporate social responsibility.

Discover the power of advanced technologies, such as Building Information Modeling (BIM) and digital transformation, to streamline processes, enhance collaboration, and optimize decision-making. Dive into the world of sustainable construction practices, harnessing the potential for green building and creating a positive impact on the environment.

Throughout the book, you'll find real-world case studies, providing invaluable insights and lessons learned from notable construction projects. These practical examples showcase the application of best practices and demonstrate how to overcome common challenges, ensuring you're equipped to tackle any project with confidence.

This is not just a book; it's your trusted companion, empowering you to unlock your full potential as a construction professional. Whether you're a seasoned project manager, business owner, or an aspiring industry leader, this book is a must-have resource that will elevate your skills, enhance your decision-making, and propel your career to new heights.

Take your construction projects and business ventures to the next level. Get your hands on the book today and embark on a transformative journey toward excellence and success in the dynamic world of construction.

Mastering The Construction Business: Guidelines For Success

In a rapidly evolving construction industry, standing out and achieving success requires more than just hard work. It demands a deep understanding of the intricacies and nuances that shape the business landscape. Discover the keys to unlocking the construction industry with this extraordinary guide.

This book is a comprehensive and indispensable resource that equips you with the tools, knowledge, and strategies needed to excel in the competitive world of construction business. From navigating complex legal and regulatory frameworks to implementing effective financial management practices, this guide covers every aspect of the business with unparalleled depth and insight.

Learn how to conduct meticulous market analysis, identify target markets, and develop winning business plans that set you apart from the competition. Explore the art of project management, from initiation to closure, and gain invaluable insights on resource allocation, scheduling, and effective team building.

With a sharp focus on risk management, you'll discover how to identify potential pitfalls, implement robust risk mitigation strategies, and safeguard your projects and investments. Harness the power of technology and innovation to streamline operations, improve efficiency, and stay ahead of the curve in this rapidly advancing digital era.

Embrace the principles of sustainability and green building to not only meet environmental demands but also create a competitive advantage in the market. Gain a deep understanding of ethical practices and corporate social responsibility, ensuring your business operates with integrity and builds strong, long-lasting relationships.

Designed to empower professionals at all levels of the construction industry, this guide goes beyond theory, providing practical examples, case studies, and actionable insights that will transform your approach to the construction business.

Construction Health And Safety Fundamentals

In the fast-paced world of construction, ensuring the health and safety of workers is paramount. This book is an indispensable guide that lays the foundation for creating a culture of safety and excellence in the construction industry.

Authored by a seasoned expert in construction health and safety, the book offers a comprehensive exploration of the fundamental principles, strategies, and best practices that underpin effective safety management. It covers a wide range of topics, including safe operation of heavy machinery, crane and hoist safety, rigging and lifting operations, power tool safety, electrical safety, fire safety and emergency response, working at heights and fall protection, confined space entry and rescue, and much more.

Designed as a practical resource, each chapter provides clear explanations of key concepts and actionable insights. One of the distinguishing features of this book is its comprehensive approach to emerging trends and technologies in construction health and safety. It explores innovative solutions such as wearable technologies, virtual reality training, and predictive analytics, empowering readers to stay ahead of the curve and leverage cutting-edge tools for enhanced safety outcomes.

Whether you are a construction professional, safety manager, project supervisor, or worker looking to enhance your knowledge and skills in construction health and safety, this book is your go-to resource. Don't compromise on safety! Together, we can build a safer and more prosperous future for the construction industry.

Construction Management Blueprint: A

Comprehensive Guide To Successful Project Delivery

Are you ready to take your construction management skills to the next level? Look no further! This book is the ultimate guidebook that will equip you with the knowledge and tools you need to excel in the complex world of construction projects.

This comprehensive book covers every aspect of construction management, from project initiation to post-construction evaluation. Written by industry experts with decades of experience, it provides a wealth of practical insights, real-life case studies, and invaluable tips for success.

Inside these pages, you'll find:
A step-by-step guide to project initiation and feasibility studies, helping you identify objectives, assess market demand, and engage stakeholders effectively.
In-depth coverage of project planning and design, including goal setting, scope definition, work breakdown structures, and sustainable design principles.
Extensive discussions on cost estimation techniques, budgeting, resource allocation, value engineering, and contract pricing and negotiation strategies.
Detailed insights into construction scheduling, resource procurement, site layout and logistics, and risk management, ensuring smooth project execution.
Thorough examinations of quality assurance and control, materials testing and inspection, occupational health and safety practices, and risk mitigation strategies.
Expert guidance on commissioning and handover, facility documentation, owner training, and maintenance for long-term sustainability.
Essential information on post-construction evaluation, continuous improvement, professional development, and knowledge management in construction management.

The book is not just a theoretical guide; it's a practical companion for construction professionals, project managers, and students looking to enhance their skills and achieve outstanding results. Its clear and concise language, combined with visually engaging diagrams and insightful case studies, makes it an enjoyable and accessible read for both industry veterans and newcomers.

Whether you're involved in residential, commercial, or infrastructure projects, the book will empower you to navigate the complexities of the construction industry with confidence and achieve excellence in every aspect of your work.

Don't miss out on this opportunity to unlock your full potential as a construction management professional. Get your copy of the book today and embark on a journey toward unparalleled success in the dynamic world of construction projects.

Construction Management Fundamentals: A Handbook For Construction Students, Academics, And Practitioners

Are you seeking a comprehensive guide to mastering the fundamentals of construction management? Look no further than this book! This handbook offers a clear and concise overview of the principles, practices, and techniques of construction management. It covers the concepts essential for a successful career in construction management, including project planning, scheduling, and construction contracts. The book is designed to be an invaluable resource for construction students, academics, and practitioners alike.

Whether you're a seasoned professional looking to refresh your skills or a student just starting out in the field, this handbook is an indispensable resource that will help you excel in your career.

Get your copy today and take your construction management skills to the next level!

Building Maintenance Guidelines: A Complete Manual

This book is an essential resource for anyone responsible for maintaining and preserving the integrity of a building. It covers several aspects of building maintenance, from electrical systems and HVAC systems to roofing, plumbing, and structural components. It provides clear, step-by-step instructions on how to perform routine maintenance tasks. It also includes information on how to identify potential problems, such as water damage, mold growth, and insect infestations, and provides guidance on how to address these issues. In addition to its practical information, the book also includes important information on energy efficiency and sustainability.

With its clear, easy-to-follow language, the book is an invaluable resource for anyone looking to keep their building in optimal condition.

www.ingramcontent.com/pod-product-compliance
Lightning Source LLC
Chambersburg PA
CBHW072156290526
45794CB00004B/1540